Mainstreaming Renewable Energy in the 21st Century

JANET L. SAWIN

Thomas Prugh, *Editor*

WORLDWATCH PAPER 169

May 2004

THE WORLDWATCH INSTITUTE is an independent research organization that works for an environmentally sustainable and socially just society, in which the needs of all people are met without threatening the health of the natural environment or the well-being of future generations. By providing compelling, accessible, and fact-based analysis of critical global issues, Worldwatch informs people around the world about the complex interactions among people, nature, and economies. Worldwatch focuses on the underlying causes of and practical solutions to the world's problems, in order to inspire people to demand new policies, investment patterns, and lifestyle choices.

FINANCIAL SUPPORT for the Institute is provided by the Aria Foundation, the Deutsche Gesellschaft für Technische Zusammenarbeit (German Society for Technical Co-operation), The William and Flora Hewlett Foundation, The Frances Lear Foundation, the Steven C. Leuthold Family Foundation, the Merck Family Fund, The Overbrook Foundation, the V. Kann Rasmussen Foundation, the A. Frank and Dorothy B. Rothschild Fund, The Shared Earth Foundation, The Shenandoah Foundation, The Summit Fund of Washington, the UN Population Fund, the Wallace Genetic Foundation, Inc., the Wallace Global Fund, the Johanette Wallerstein Institute, and The Winslow Foundation. The Institute also receives financial support from its Council of Sponsors members Adam and Rachel Albright, Tom and Cathy Crain, John and Laurie McBride, and Tim and Wren Wirth, and from the many other friends of Worldwatch.

THE WORLDWATCH PAPERS provide in-depth, quantitative, and qualitative analysis of the major issues affecting prospects for a sustainable society. The Papers are written by members of the Worldwatch Institute research staff or outside specialists and are reviewed by experts unaffiliated with Worldwatch. They have been used as concise and authoritative references by governments, nongovernmental organizations, and educational institutions worldwide. For a partial list of available Worldwatch Papers, go online to www.worldwatch.org/pubs/paper.

Contents

Acknowledgments: I would like to thank Christian Kjaer, Corin Millais, and Christine Real de Azua for their helpful comments on a draft of this paper, and David Milborrow and Eric Martinot for responding to my many e-mails for information. I am also grateful to Paul Maycock, who gave generously of his time to bring me up to date on developments and statistics in the photovoltaics industry, and to Birger Madsen for providing data and other insights into developments in the global wind power industry. Thank you also to Anand Rao and Miquel Muñoz for their research assistance.

Within the Institute, this work was greatly strengthened by input from Chris Flavin, Worldwatch's president and long-time renewable energy expert, and editing assistance provided by Thomas Prugh. Thank you also to Lyle Rosbotham, the Institute's art director, for putting my words into their final form, and to Susan Finkelpearl and Heather Wilson for all of their help with outreach.

Janet L. Sawin is a Research Associate at the Worldwatch Institute, where she leads the energy/climate change team. She has authored two chapters in the Institute's *State of the World* annual, as well as pieces for *World Watch* magazine and *Vital Signs*. Janet recently completed the background paper on "National Policy Instruments" for the International Conference for Renewable Energies (Renewables 2004) to be held in Bonn, Germany, in June 2004. She holds a B.A. in economics from Carleton College, and an M.A. and Ph.D. in international energy and environmental policy from the Fletcher School of Law and Diplomacy.

SUMMARY

Wind and solar power are the world's fastest-growing energy sources, with capacity expanding at double-digit rates every year over the past decade. Globally, wind power already generates electricity equal to that used by 19 million European households. In 2003, an estimated $20.3 billion—about one-sixth of total global investment in power generation equipment—were invested in "new renewables" (all renewable energy sources except large-scale hydropower and traditional biomass). The effects of this rapid growth include impressive technology advances, dramatic cost reductions, and an increase in political support for renewable energy around the world.

These developments occur against a backdrop of rapidly rising demand for energy, as well as growing concerns about the security of energy supplies and the environmental and health dangers associated with the burning of fossil fuels. Indeed, the need for new, sustainable sources of energy has never been greater. Although new renewables currently meet only 2 percent of global energy demand, the technical potential of these inexhaustible and relatively benign energy sources far exceeds total energy use.

A mere six countries—Denmark, Germany, India, Japan, Spain, and the United States—account for about 80 percent of global photovoltaic (PV) and wind power capacity. In all cases, the advancement of renewables has been spurred by strong government policies designed to nurture nascent energy industries and to create demand for these technologies, often in markets dominated by mature, heavily subsidized fossil fuels and nuclear power.

Experience shows that renewable energy can advance dramatically worldwide if governments enact the right mix of policies. Among the key policy lessons:

• Access to the market must be ensured. Pricing laws have proved most successful to date at creating markets, while also encouraging steady industry growth and private sector investment in R&D, and offering ease of financing. Quota systems (such as the renewable portfolio standards established in several U.S. states) have also been useful.

• Financial incentives (including tax credits, rebates, payments, and low-interest loans) are also important for encouraging investment in renewables by reducing investors' risks and compensating for high initial capital costs. Subsidies should be phased out as costs decline.

• Education and information dissemination are necessary to apprise potential investors and customers about the potential of renewables, dispel myths, and ensure that trained workers are available to produce, install, and maintain renewable energy equipment.

• Public participation and ownership in the renewables development process increase political support and the likelihood of success.

• Industry standards and permitting help prevent inferior hardware from entering the marketplace and eroding investor and customer confidence, while also addressing potential sources of opposition such as noise and visual impacts.

Governments also must rethink their relationships to the conventional energy industry. Reducing or eliminating the hundreds of billions of dollars in annual subsidies, incorporating all costs into the price of energy, and shifting government purchases from conventional to renewable energies would help to level the playing field for renewable technologies. Finally, policies enacted to advance the development and use of renewables must be consistent and long-term to avoid boom-and-bust environments that shake investor confidence and choke off the supply of capital.

Introduction

Renewable energy is poised for a global takeoff. Over the past decade, the installed capacity of solar power has increased seven-fold, and wind energy capacity has grown by more than a factor of 13. These 10-year annual growth rates (of 22 and 30 percent, respectively) are closer to the realm of computers and telecommunications than the single-digit growth rates common in today's energy economies. And their impact could be revolutionary. The immediate effects include rapidly declining costs, impressive technology advances, and growing economic power and broad-based political support, which in turn are leading to further policy reforms and even faster growth.

Those in the mainstream energy sectors tend to dismiss rapid growth in what they view as tiny industries. This thinking mirrors the attitude of IBM toward Microsoft in the early 1980s. Such high growth rates can rapidly vault a new industry from insignificance to market dominance and thus radically transform the status quo. The conventional wisdom is that the high growth rates will quickly decelerate. Yet the global capacity of both wind and solar photovoltaics (PV) has grown faster over the past five years than in the previous five. And in fact these industries are already far from tiny. For example, today's worldwide wind capacity is sufficient to power the equivalent of 19 million European households.

Although "new renewables" (which exclude large-scale hydropower and traditional biomass) still represent a modest 2-percent share of global energy use, and wind and solar represent less than 1 percent, these new energy sources are large enough to command attention in the marketplace. The estimated $20.3 billion spent on renewable energy development in 2003 was roughly one-sixth of total world investment in power generation equipment.[1*] And these industries are attracting some of the largest players in the world energy market, including BP, Royal Dutch/Shell, and General Electric.

The vast potential of these energy sources is shown by the fact that the past decade's growth in renewable energy has taken place mostly in six countries, which represent roughly 80 percent of the world's generation of wind and solar power. Unlike the markets for oil or coal, the dominant roles of Denmark, Germany, India, Japan, Spain, and the United States in renewables do not reflect a fortunate accident of geography and resource availability. They are instead the product of conscious policy decisions that have created demand for these technologies, including access to the electric grid at attractive prices, low-cost financing, tax incentives and other subsidies, standards, education, and stakeholder involvement. Public research and development investments are also important, but it is only by creating markets that the technological development, learning, and economies of scale in production can develop to further advance renewables and reduce their costs. The costs of these policies have been relatively minor compared to the leverage they have provided, spurring billions of dollars' worth of research and development and capital investment by the private sector.

These six countries have shown that it is possible to create vibrant markets for renewable energy and to do so rapidly. New laws to promote renewables are being introduced almost continually at the state and national levels worldwide. If more countries continue to board the renewable energy bandwagon, renewables could reach a tipping point that propels them toward dominance of the global energy system—much as oil passed a similar threshold a century ago—and provide humanity with a cleaner, safer, healthier, and more equitable world.

---

*Endnotes are grouped by section and begin on page 53.

For renewable energy to make a significant contribution to economic development, job creation, reduced fossil fuel dependence, improved human health, and lower greenhouse gas emissions, it is essential to improve the efficiency of the technologies, reduce their costs, and develop mature, self-sustaining industries to manufacture, install, and maintain renewable energy systems. The goal must be to establish the conditions for sustained and profitable industries. These in turn will boost renewable energy capacity and generation, and will drive down costs. Viable, clear, and long-term government commitments are critical to this end, along with policies that create markets and ensure a fair rate of return for investors.

The need for new energy sources has never been greater. Energy use is rising rapidly everywhere but particularly in the developing world, where up to 2 billion people still lack access to electricity and other modern energy services, and average per-person energy consumption is far below that in the industrial world.[2] For most developing countries that lack fossil fuels but are rich in renewable and human labor resources, renewable energy is a perfect match, making it possible to create millions of jobs while reducing the foreign exchange burden of imported fuels. The same holds true for much of the industrial world as well, where renewables can meet rising demand and replace obsolete systems.

On the other hand, if the world continues down the track of business as usual, it faces a fast-approaching train wreck. Oil is being consumed at ever more rapid rates, and the peak in world oil production could be less than a generation away. Not only are conventional fuels insufficient to meet rising energy needs through this century, but they also impose unacceptable economic, health, social, and security costs. For instance, the steady rise of atmospheric carbon dioxide levels—and the consequent risk of climate change, whether gradual or abrupt—is now receiving the attention of everyone from urban planners to Pentagon strategists.

Although a transition to renewable energy will require considerable upfront investment, numerous studies conclude that it would be cheaper over the long term, while also providing tremendous social, economic, security, and environmental advantages. Just as the United States dominated the petroleum economy of the last century, countries that invest in renewable energy technologies early on will be in a strong position to reap the economic rewards of a rapidly growing new sector.*

The Approaching Train Wreck— and How To Avoid It

During the past year, Shanghai's gleaming shopping malls have gone for hours without heat on winter days, while children study by candlelight and factories shut down for lack of power.[1] The lights are out across much of China because energy supply cannot keep up with rapidly rising energy demand, driven by extraordinary economic growth. Simultaneous shortages of oil, electricity, and coal have sparked concerns about an impending energy crisis and, ironically, are slowing further economic expansion.[2] At the same time, the World Bank estimates that the environmental and health costs of air pollution in China, due primarily to coal burning, could total 13 percent of China's gross domestic product (GDP) by 2020.[3]

China's electricity use has tripled since 1990.[4] In 2003 alone, power demand jumped 15 percent, and oil consumption increased more than 10 percent.[5] A decade ago, it was a net exporter of oil; in 2003, due primarily to a dramatic rise

---

* In June 2004, the German government will host the first major intergovernmental conference on renewable energy since the 1981 UN Conference on New and Renewable Sources of Energy in Nairobi. Major issues to be discussed will include barriers to the development and diffusion of renewable energy technologies, policy instruments for advancing their use, and financing to accelerate development. The purpose is to develop an international action plan with voluntary national and regional targets aimed at substantially increasing the global share of energy from renewable sources. The International Conference for Renewable Energies in Bonn will offer a historic opportunity for nations to unite toward the common goal of a more sustainable energy future, and to work together to bring renewables into the mainstream during the 21st century.

in private car ownership, China passed Japan to become the second largest consumer after the United States. Long a major exporter of coal, China could become a major importer within four years.[6]

And China is not alone. More and more people in the global South are using as much energy on average as people in the North do, and studies suggest that their incomes are rising faster than those in the industrial world.[7] Demand for energy will continue to rise as people in developing countries increasingly adopt the transportation systems, diets, and lifestyles of consumers in the world's richest nations. The International Energy Agency (IEA) projects that, between 2000 and 2030, global energy consumption will increase 66 percent and electricity use could double.[8] The largest share of this growth will likely occur in the developing world.

New conventional power plants will come on line in China by 2006, easing current shortages. But they will be only temporary fixes for an emerging challenge that developing and industrial nations alike must soon confront: how to satisfy the world's voracious and growing appetite for energy, which is relentlessly increasing the pressure on non-renewable resources, public health and welfare, international stability, and the natural environment.

Even at current global consumption rates, many analysts predict that world oil production will peak before 2020, and while the world will technically never run out, fossil fuels will become increasingly difficult and expensive to extract.[9] According to Harry Shimp, former president of BP Solar, "In 20 to 25 years the reserves of liquid hydrocarbons are beginning to go down so we have this window of time to convert over to renewables."[10] Of greater concern to many, however, is not when or if economically recoverable fossil fuel reserves will be depleted, but the fact that the world cannot afford to use all the conventional energy resources that remain.

Worldwide, there is a growing realization that climate change, caused primarily by the burning of fossil fuels, is a more serious threat to the international community than terrorism and that it "remains the most important global challenge to humanity."[11] The Intergovernmental Panel on Climate Change (IPCC), a body of 2,000 scientists and economists that advises the United Nations on climate change, has concluded that global carbon dioxide (CO_2) emissions must be reduced at least 70 percent over the next 100 years to stabilize atmospheric CO_2 concentrations at 450 parts per million, which would be 60 percent higher than pre-industrial levels.[12] There is evidence that effects of global warming are already being felt worldwide, as weather-related disasters grow more frequent and costly and associated death rates rise.[13] The sooner societies begin to reduce their emissions, the lower will be the impacts and associated costs of both climate change and emissions reductions.

Other environmental costs of conventional energy production and use include the damage wrought by resource extraction; air, soil, and water pollution; acid rain; and biodiversity loss. Conventional energy requires vast quantities of fresh water. Mining and drilling affect the way of life and very existence of indigenous peoples worldwide. Urban air pollution from burning fossil fuels is responsible for hundreds of thousands of premature deaths each year around the world.[14] In the European Union, the environmental and health costs associated with conventional energy (and not incorporated into energy prices) are estimated to equal 1–2 percent of the EU's GDP, excluding costs associated with climate change.[15] (See Table 1.)

The direct economic and security costs associated with conventional energy are also substantial. Nuclear power is one of the most expensive means of generating electricity, even without accounting for the risks of nuclear accidents, waste, and weapons proliferation. All conventional power plants face risks of conflict, sabotage, accidents, or even disruption of fuel supply. And massive and costly power blackouts are difficult to avoid in highly centralized systems of production and distribution based on fossil fuels and nuclear power.

Political, economic, and military conflicts over limited energy resources will intensify as global demand increases. Similarly, as demand rises and supply becomes further concentrated in the world's unstable, resource-rich regions, the prices of oil

TABLE 1

Costs of Electricity With and Without External Costs

Electricity Source	Generating Costs	External Costs	Total Costs
		(U.S. cents per kilowatt-hour)	
Coal/lignite	4.3–4.8	2.3–16.9	6.6–21.7
Natural gas (new)	3.4–5.0	1.1–4.5	4.5–9.5
Nuclear	10–14	0.2–0.8	10.2–14.8
Biomass	7–9	0.2–3.4	7.2–12.4
Hydropower	2.4–7.7	0–1.1	2.4–8.8
Photovoltaics	24–48	0.7	24.7–48.7
Wind	3–5	0.1–0.3	3.1–5.3

Notes: Generating costs are for the United States and/or Europe. External costs are environmental and health costs for 15 countries in Europe, and are converted to U.S. cents from eurocents at the 2003 average exchange rate of US$1= € 0.8854.
Sources: See Endnote 15 for this section.

and gas will become increasingly erratic, affecting the stability of economies around the world. In fact, there is increasing evidence that a rise in fossil fuel prices or volatility leads to economic decline, even global recessions.[16] Further, the economic costs of relying on imported fuels are extremely high. African countries, for example, spend an estimated 80 percent of their export earnings on imported oil.[17] Conversely, the benefits of reducing imports can be significant. Brazil's 27-year-old ethanol program, which displaces about 220,000 barrels of oil daily, has saved Brazil more than $52 billion in avoided fuel imports, many times the total investments in ethanol production.[18]

Change is never easy, and there are strong forces (including politically powerful industries) acting to maintain the status quo. Yet while the world remains sharply divided over what kind of energy future must lie ahead, political support for renewable energy is on the rise as strong new legislation opens markets for renewable energy in a rapidly growing list of countries. Many nations view renewable energy as not only a credible alternative to fossil fuels, but also a necessity to meet growing energy needs without sacrificing quality of life, human health, the natural environment, and national security.

New renewable resources provide only a small share of

global energy production today.[19] (See Figures 1 and 2.) Yet renewable energy technologies have the potential to meet world energy demand many times over and are ready for use on a large scale.[20] (See Table 2, page 16.) Renewable energy can generate electricity, heat and cool space, perform mechanical work such as water pumping, and produce fuels—in other words, everything that conventional energy does.

Moreover, the advantages of shifting from conventional energy to renewable energy are numerous and compelling. Renewable technologies impose significantly lower social, environmental, and health costs than do conventional fuels and technologies. They are generally domestic, pose far fewer fuel and transport hazards, and are much less vulnerable to terrorist attack. Generating power locally with solar or wind energy, for example, reduces or eliminates transmission and distribution losses, which range from 4 to 7 percent in industrial nations to more than 40 percent in parts of the developing world.[21] Renewable technologies can be installed rapidly and in dispersed small- or large-scale applications—getting power quickly to areas where it is urgently needed, delaying or avoiding investment in expensive new electric plants or power lines, reducing investment risk, and promoting economic development. All renewables except biomass energy avoid fuel costs and the risks associated with future fuel price fluctuations. It has been estimated that investments required over a 10-year period to make renewables competitive worldwide within two decades would be far lower than the economic costs of a single 10-percent increase in oil prices, and would be modest in comparison with existing flows for energy sectors worldwide.[22]

Renewables can provide reliable power for businesses in developing countries like China and India where power cuts are common. India's former minister for nonconventional energy sources, M. Kannappan, has declared that renewables have "enormous potential to meet the growing requirements of the increasing populations of the developing world, whilst offering sustainable solutions to the threat of global climate change."[23] Developing countries that invest

FIGURE 1

World Energy Use by Source, 2000

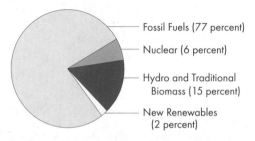

Fossil Fuels (77 percent)

Nuclear (6 percent)

Hydro and Traditional Biomass (15 percent)

New Renewables (2 percent)

FIGURE 2

World Electricity Generation by Type, 2001

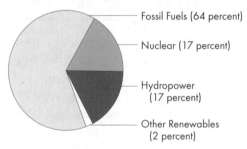

Fossil Fuels (64 percent)

Nuclear (17 percent)

Hydropower (17 percent)

Other Renewables (2 percent)

Sources, Figures 1 and 2: See Endnote 19 for this section.

in renewables will discover that they are energy-rich—that they can leapfrog over the dirty technologies relied on by early industrializing countries and can develop their economies with clean, domestic, secure sources of energy that avoid long-term and costly imports.

Further, "renewables [are] not just about energy and the environment but also about manufacturing and jobs." This ringing endorsement came from U.K. Energy Minister Brian Wilson in July 2002, after the commissioning of a new 30-megawatt wind farm in Argyll, Scotland. The Kintyre Peninsula of Argyll once thrived on its fisheries, whiskey production, and textile manufacturing. But these traditional sources of employment are in decline, and now wind power is breathing new life into the region's economy, generating power for 25,000 homes and producing new jobs.[24]

TABLE 2

Global Renewable Resource Base (Exajoules/year)

Resource	Current Use	Technical Potential
Hydropower	10	50
Biomass	50	>250
Solar	0.2	>1,600
Wind	0.2	600
Geothermal	2	5,000
Total Renewables	62.4	>7,500
Total Global Energy Use, 2000	422.4	–

Notes: Data are for late 1990s. Total global energy use includes traditional biomass. Technical potential is based on available technologies and will increase as technologies improve. *Sources: See Endnote 20 for this section.*

Around the world, using renewables stimulates local economies by attracting investment and tourists (and their money) and by creating employment. Many of the jobs are high-wage and high-tech, and require a range of skills, often in rural or economically depressed areas.[25] A recent study concluded that increasing the use of renewable energy technologies in California would create four times more jobs than continued operation of natural gas plants, while keeping billions of dollars in California that otherwise would go for out-of-state power purchases.[26]

Many of the components, if not entire systems, for solar homes, wind farms, and other renewable technologies are now manufactured or assembled in developing countries, creating local jobs, reducing costs, and keeping capital investments at home. For example, China and India have both developed domestic wind-turbine manufacturing industries, with Indian firms producing about 500 MW of turbines annually for domestic use and export.[27]

The many advantages of renewables led the Task Force on Renewable Energy of the Group of Eight (G8) industrial countries to conclude in 2001 that "though there will be a higher cost in the first decades, measured solely in terms of the costs so far reflected in the market, successfully promoting renewables over the period to 2030 will prove less expensive than taking a 'business as usual' approach within any realistic range of discount rates."[28]

Technology and Market Development

Since the 1980s, renewable technologies have improved significantly in both performance and cost, with some undergoing rates of growth and technology advancement comparable to the electronics industry. Wind and solar power are the fastest-growing energy sources in the world.[1] (See Figure 3, page 18.) By some estimates, new renewables already account for well over 100,000 megawatts (MW) of grid-connected electric capacity. Globally, new renewable energy supplies the equivalent of the residential electricity needs of more than 300 million people.[2]

In 1999, the International Energy Agency noted that "the world is in the early stages of an inevitable transition to a sustainable energy system that will be largely dependent on renewable resources."[3] This is a bold statement for an organization that represents North America, Europe, and Japan—areas that depend so heavily on fossil fuels. But it seems logical, given the many problems associated with the use of conventional energy and the tremendous surge in renewable energy investments over recent years.

Global investment in renewable energy exceeded $20.3 billion in 2003, and cumulative investments totaled at least $100 billion between 1995 and 2003.[4] Markets for new renewable energy are expected to approach $85 billion annually within the next decade.[5] The technical progress of many renewable technologies has been faster than anticipated even a few years ago, and this trend is expected to continue. While costs are still a concern with some technologies, they are falling rapidly due to technological advances, automated manufacturing, economies of scale through increased production volumes, and learning by doing.[6]

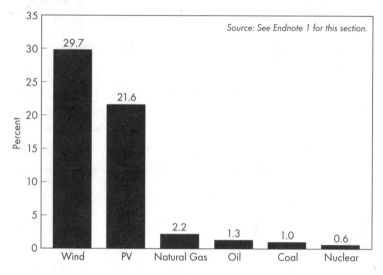

FIGURE 3

Average Annual Increase in Installed Capacity, 1993–2003

Source: See Endnote 1 for this section.

Solar and wind are the best-known renewables, but inexhaustible energy supplies are also offered by biomass, geothermal, hydropower, ocean energy (from tides, currents, and waves), and ocean thermal energy. The remainder of this paper focuses on wind power and photovoltaics for electricity generation because they are the fastest-growing renewables, they share the challenges of being intermittent and having high up-front capital costs, solar and wind resources are nearly ubiquitous, and they have the greatest potential for helping all countries to achieve a more sustainable energy future.

During the past two decades, wind energy technology has evolved to the point where it can compete with conventional forms of power generation at good sites. Costs have declined 12–18 percent for each doubling of global capacity.[7] As a result, the average cost of wind-generated electricity has fallen from about 46 cents per kilowatthour (kWh) in 1980 to 3–5 cents/kWh at good wind sites today.*[8] Costs vary from one

* 1980 costs are for United States only. All costs are in 2003 U.S. dollars.

location to the next due primarily to variations in wind speed and different institutional frameworks and interest rates. By 2010, onshore wind generation costs will likely be lower than natural gas costs, and offshore wind costs could fall by 25 percent.[9] As costs fall, it will become economical to site turbines in regions with lower wind speeds, increasing the global potential for wind-generated electricity.

The main trends in wind technology development are toward lighter and more flexible blades, variable speed operation, direct-drive generators, and taller machines with greater capacity.[10] The average turbine size has increased from 100–200 kilowatts (kW) in the early 1990s to more than 1,200 kW today, making it possible to produce more power with fewer machines.[11] (One 1,200 kW machine can provide the electricity needed by about 720 European homes.) Larger machines are available for use on land, and turbines with capacity ratings as high as 5,000 kW (5 MW) are being manufactured for use offshore.[12] Small wind machines that can be installed close to the point of demand (atop buildings, for example) are also under development.[13] Advances in turbine technology and power electronics, along with a better understanding of siting needs and wind energy resources, have combined to extend the lifetime of today's wind turbines, improve performance, and reduce costs.[14] (See Sidebar 1, page 20.)

Global wind capacity has grown at an average annual rate of nearly 30 percent during the past decade.[15] (See Figure 4, page 21.) An estimated 8,250 MW of wind capacity were added worldwide in 2003, bringing the total to nearly 40,290 MW—enough to provide power to more than 19 million average European households.[16] It has taken 25 years to reach this total; if the industry's projections hold true, another 110,000 MW could be added in only nine years.[17] Wind is now generating electricity in at least 48 countries.[18] However, Europe accounts for more than 70 percent of total global capacity, and most of these installations are in only three countries (Germany, Denmark, and Spain) where onshore markets have begun to peak due to some market saturation, and offshore projects have experienced slow starts. But the overall health of the industry is good, and

SIDEBAR 1

Examples of Advances in Wind Technology

The U.S. Department of Energy's Sandia National Laboratory is developing turbines with lighter, larger, slower-rotating blades to reduce costs and produce power at less windy sites. These next-generation turbines could expand wind development potential as much as 20 times.

Mathematical climate models have been developed in Germany and Denmark to predict wind resources 24–36 hours in advance with reasonable accuracy. This will be important for managing wind power capacity as it reaches a high percentage of the total electric system.

Vestas now equips offshore turbines with sensors to detect wear and tear on components, along with backup systems to cope with electronic system power failures.

Sources: See Endnote 14 for this section.

significant projects are in the pipeline in the United Kingdom and other countries that could become wind powerhouses of the future.[19] Global sales of wind power worldwide exceeded $9 billion in 2003 and are predicted to reach $49 billion annually by 2012.[20] It is estimated that more than 100,000 people are now employed in the wind industry worldwide.[21]

The majority of turbines operating today are on land, but new markets are opening for wind power offshore, mainly in Europe, because the resource is huge and wind speeds at sea are considerably higher and more consistent. (Stronger winds generate more electricity, while consistency reduces wear and tear on machines.) By the end of 2003, turbines with a combined capacity of 529 MW were spinning offshore, all of them in Europe, with an additional 8,600 MW planned for construction through 2008.[22]

Resource analysis shows that onshore wind resources could supply more than four times as much electricity as is now consumed worldwide. Offshore resources are substantial as well. While some of that potential is too costly to exploit over the near term, the promise of large amounts of wind power at competitive prices is enormous.[23]

As with all energy technologies, there are disadvantages

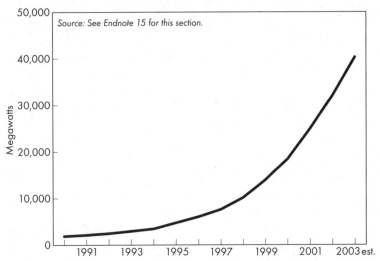

FIGURE 4

Cumulative Global Wind Capacity, 1990–2003

Source: See Endnote 15 for this section.

associated with wind power. The environmental factor that has caused the most controversy and concern is bird mortality. This is a site-specific problem, however, and it is far less significant than other hazards to birds, such as vehicles, buildings, cell phone towers, and (the primary threat) habitat loss. Progress has been made in reducing bird strikes through the use of painted blades, slower rotational speeds, tubular turbine towers, and careful siting of projects.[24]

Both wind and sun are intermittent resources; they cannot be turned on and off as needed. But there is no guarantee that any resource will be available when it is required, and utilities must have backup power for generation every day. Assessments in Europe and the United States have concluded that intermittent sources can account for up to 20 percent of an electric system's generation without posing technical problems; higher levels might demand minor changes in operational practices.[25] Wind power's contribution already exceeds 20 percent in regions of Germany, Denmark, and Spain, and distributed generation—for example, the use of solar panels on rooftops, or clusters of turbines along the path of a power line—

can improve electric system reliability.

The challenges posed by intermittency are not of immediate concern in most countries, where the share of electricity from the sun and wind are far from 20 percent. Where necessary, they will be addressed via hybrid systems (for example, a mix of wind- and hydropower), improvements in wind forecasting technology, and further development of storage technologies.[26] New storage technologies could also help tap renewable resources that are far from demand centers. Furthermore, what is most significant is the cost of electricity generated. Wind power costs continue to fall, and at good locations are competitive with all conventional technologies. Solar PVs are likely to see dramatic cost reductions, and they produce power in the middle of hot summer days when demand is greatest and electricity costs are highest. High-tech solutions now under development, such as "smart grids" (which use advanced computer controls to enable more efficient, resilient, and safe distribution of power) can provide renewable energy with easier access to energy markets while improving the performance and cost of renewable energy systems.[27]

According to the U.S. National Renewable Energy Laboratory, PV technologies have the "potential to become one of the world's most important industries." The potential PV market is enormous, ranging from consumer products (such as calculators and watches) and remote stand-alone systems for electricity and water pumping to grid-connected systems on buildings and large-scale power plants.[28] Today, 60–70 percent of solar electric power is fed to electric grids.[29]

Each year the sun delivers to Earth more than 10,000 times the energy that humans currently use.[30] While PV systems account for a small share of global electricity generation, they have undergone dramatic growth over the past decade. Since 1993, global PV production has increased at an average annual rate exceeding 28 percent, and growth rates have risen almost every year. It took nearly 30 years, until 1999, for the world to produce its first gigawatt (GW) of PV capacity; by the end of 2003, this total had tripled.[31] (See Figure 5.) The PV industry generated sales worth more than $5.2 billion in

FIGURE 5

Cumulative Global Photovoltaic Production, 1990–2003

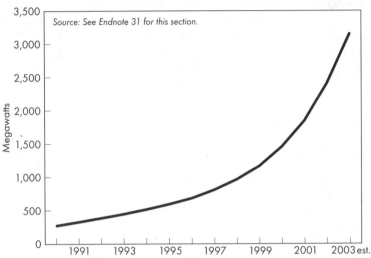

Source: See Endnote 31 for this section.

2003, and provides tens of thousands of jobs.[32] It is projected that the industry will surpass $27.5 billion in annual sales (including components and installation) by 2012.[33] More than a million households in the developing world now have electricity for the first time from PVs, while well over 100,000 households in industrial countries supplement their utility power with PV systems.[34]

PV technology has advanced significantly over the years, primarily through incremental improvements. Crystalline silicon cells and modules, which now dominate the market, have achieved commercial efficiencies of 12 to 15 percent; they could reach 20 percent by 2010 and 30 percent or higher in the longer term.[35] Sanyo has developed a technology combining amorphous silicon and crystalline silicon that has achieved module efficiencies of 17 to 19 percent, and other major players are developing single-crystal silicon modules with similar efficiency rates. As a result, even the typically small rooftops in Japan can generate as much power as the average Japanese household consumes, allowing for "zero-energy" homes.[36] The industry is now highly competitive with a host of technologies, including multi-crystalline and triple-junction cells, amorphous thin films, polymers, dyes, concentrating lens PV hybrids, and nanotechnology.[37] (See Sidebar 2.)

Solar cell production is concentrated in Japan, Europe, and the United States, but there are growing markets and manufacturing bases in developing countries as well, including China and India. Global PV output is expected to increase by 25 percent per year through 2010.[38] As larger factories come into operation, manufacturers can increase the degree of automation.

Such evolving industrial processes, along with technological advances and economies of scale, have led already to significant cost reductions. Since 1976, costs have dropped 20 percent for every doubling of installed PV capacity, or about 5 percent annually.[39] Module prices have declined from about $30 per watt ($/W) in 1975 to below $4/W today, with some bulk purchases costing less than $3/W.[40] Globally, total system costs—including balance-of-system components such as inverters, and installation—range from a low of $5.25–5.50/W in Japan, to $6–8/W in California, to a high of $20/W for remote, off-grid systems.[41]

As with wind energy, actual generating costs are determined by capital costs (for modules, other system components, and installation), interest rates, and the available resource. Generating costs worldwide now range from $0.11 to $1.00 per kWh, which is still extremely high at the upper end, and cost remains the primary barrier to more widespread use.[42] Yet PVs are the cheapest option for many remote or off-grid functions. When used for building facades, they can be cheaper than other materials such as marble or granite, with the added advantage of producing electricity.[43] And PV systems are now competitive on-grid at all times in Japan, and at peak demand times in California, where government policies and private investments have led to reduced costs through economies of scale in production and experience with installation.[44] Around the world, companies are racing to create future generations of products to make PVs cost-competitive for on-grid use elsewhere as well. Many manufacturers aim for a module price of $1 per peak watt.[45]

SIDEBAR 2

The Solar Race

The U.S. National Renewable Energy Laboratory (NREL) and Spectrolabs have developed a Triple-Junction Terrestrial Concentrator Solar Cell that is 34 percent efficient and can be manufactured for less than $1 per watt, according to NREL.

"Spheral Solar" technology developed in Canada will bond tiny silicon beads into an aluminum foil, allowing for flexible, lightweight, dramatically cheaper solar cells for a broad range of new PV uses. Plans were announced in late 2003 to commercialize the technology and build a 20-MW production facility.

U.S. Evergreen Solar, Inc. has successfully produced a prototype technology that enables the growth of four silicon ribbons from one furnace. String ribbon technology can yield more than twice as many solar cells per unit of silicon than conventional methods, reducing costs and waste.

German solar cell manufacturer Sunways recently released "Solar Blinds," a product that can protect buildings against bad weather, sun, and burglary while also producing electricity.

Sources: See Endnote 37 for this section.

Costs have already declined faster than many believed possible, using existing technologies.[46] Sharp reduced per-unit costs 30–35 percent by scaling up to a 200 MW manufacturing plant that allowed for increased automation and bulk purchases of inputs such as glass.[47] Future PV cost reductions are expected to occur primarily through continued incremental improvements in materials and module efficiency, reduced costs and increased lifetime of balance-of-system components, experience with installation, and economies of scale in production.[48] Experts believe that on-grid rooftop systems could be competitive with conventional generation in the United States within a decade, even without incentives.[49]

In addition to cost, one of the primary concerns regarding PV's ability to meet a major portion of global electricity demand is the length of time cells must operate to produce as much energy as was used to manufacture them. The energy "pay-back" period for today's modules in rooftop systems is 4–6 years, depending on the technology, with expected lifetimes of up to 30 years. Payback periods will decline as the energy efficiency of production increases.[50] PV manufacture also requires hazardous materials, including many of the chemicals and heavy metals used in the semiconductor electronics industry. There are techniques and equipment to reduce the environmental and safety risks, however, and these problems are minimal compared with those associated with conventional energy technologies.[51]

According to the International Energy Agency, buildings in industrialized nations offer enough suitable surfaces for PV to generate 15–50 percent of current electricity needs.[52] Other surfaces, such as parking lots and brownfields, could increase this share. Most on-grid PV today is used in rooftop systems, but several large, centralized PV power plants are in the works. There are plans for at least two major projects (of 5 MW and 18 MW) to be built in Germany during 2004.[53] And such projects pale in comparison to other possibilities for PV. An IEA study concluded that very-large-scale PV systems installed on 4 percent of the world's deserts could produce enough electricity annually to meet world power demand, while helping to prevent further desertification. The Gobi Desert area between western China and Mongolia could generate as much electricity as current world primary energy supply.[54]

Global markets for renewables are only just beginning a dramatic expansion, starting from relatively low levels. It is useful to point out, however, that despite increasing concerns regarding safety and high costs, it took fewer than 30 years for nuclear power to develop into an industry that supplies 17 percent of global electricity demand. The same can happen with renewable technologies. In fact, since 1993 the nuclear power industry has added only 59 percent as much capacity to the world's electric grid as the wind industry.[55] If the average annual market growth rates of PV (37 percent) and wind (26 percent) over the past five years were to continue to 2020, the world would have nearly 570,000 MW of installed solar PV capacity and more than 2 million MW of wind capacity. Wind alone could supply one-fifth the electricity projected to be used worldwide in 2020.[56] Such continued growth is unlikely, but recent industry reports have concluded that if the necessary institutional framework is put in place, it is feasible for wind to meet 12 percent of global electricity demand by 2020 and for PVs to meet 26 percent by 2040.[57]

The rapid expansion of renewable technologies over the past decade has been fueled by a handful of countries that have adopted ambitious, deliberate government policies aimed at advancing renewable energy through sustained and orderly market growth. These successful policy innovations have been the most important drivers in the advancement and diffusion of renewable technologies. By examining the policies that have succeeded over the past two decades, as well as those that have failed, we can better understand what is required to launch a global takeoff in renewables in the decade ahead.

Two Success Stories: Germany and Japan

Since the early 1990s, Germany and Japan have achieved dramatic successes with renewable energy and today lead the world in the use of wind and solar power, respectively. The common elements to their stories are long-term commitments to advancing renewable energy, effective and consistent policies, the use of gradually declining subsidies, and an emphasis not only on government R&D but also on market penetration.

When the 1990s began, Germany had virtually no renewable energy industry and seemed unlikely ever to be in the forefront of these technologies. Yet within 10 years Germany had transformed itself into a renewable energy leader. With a fraction of the potential in wind and solar power as the United States, Germany now has more than twice as much installed wind capacity (more than one-third of global capacity) and is a world PV leader as well. In the space of a decade, Germany created a new, multibillion-dollar industry and tens of thousands of new jobs.

Driven by growing public concerns about the safety of

nuclear power, the security of energy supplies, and the environmental impacts (including climate change) of energy use, the German government passed an energy law in 1990 that required utilities to purchase the electricity generated from all renewable technologies in their supply areas, and to pay a minimum price for it—at least 90 percent of the retail price, in the case of wind and solar power. The "Electricity Feed-in Law" was inspired in part by similar policies that had proved effective in neighboring Denmark. The preferential payments for renewable energy are intended to help internalize the costs of conventional energy and compensate for the benefits of renewables.[1]

This pricing law has been adjusted many times since it took effect in 1991. Most significantly, in 2000 the German Bundestag required that renewable electricity be distributed among all suppliers based on their total electricity sales, ensuring that no one region would be overly burdened. With scientific input and advice from the various renewables industries, the Bundestag established specific per-kilowatthour payments for each renewable technology, based on the real costs of generation. The tariffs are paid for 20 years, while the rate for new projects is adjusted regularly to account for changes in the marketplace and technological advances. Electric utilities also qualify for these tariffs, thus reducing utility opposition while further stimulating the renewable energy market.[2]

Soon after the first pricing law was established, farmers, small investors, and start-up manufacturers started to create a new industry from scratch, and wind energy development in Germany began a steady and dramatic surge. Some barriers to renewables remained, but as each new hurdle arose the government enacted laws or established programs to address it. Obstacles to wind, for example, included lengthy, inconsistent, and complex siting procedures. The government responded by encouraging communities to zone specific areas for wind. As of 2000, grid operators were required to connect plants at the most suitable location and pay for necessary upgrading, eliminating barriers that arose when utilities discouraged wind development through inflated connection-related charges.[3]

Germany addressed the challenge of renewables' high initial capital costs through low-interest loans offered by major banks and refinanced by the federal government.[4] Until mid-2003, the "100,000 Roofs" program provided 10-year low-interest loans for PV installation (it ended early when all targets were met). In addition, income tax credits granted for projects and equipment that meet specified standards have provided tax deductions against investments in renewable energy projects. Over the years, these credits have drawn billions of dollars to the renewables industries.[5]

In addition, the federal and state governments have funded renewable resource studies on- and off-shore, have established institutes to collect and publish data, and have advanced awareness about renewable technologies through publication of subsidies and through architectural, engineering, and other relevant vocational training programs.[6]

Of all these policies, the pricing law has had the greatest impact. It ended uncertainties regarding whether, and at what price, producers could sell electricity into the grid. It also boosted investor confidence, making it easier for even small producers to obtain bank loans and drawing money into the industries. Increased investment drove improvements in technology, advanced learning and experience, and produced economies of scale that have led to dramatic cost reductions. The average cost of manufacturing wind turbines in Germany fell 43 percent between 1990 and 2000, and the cost of total PV systems declined 39 percent between 1992 and 2002.[7]

Not surprisingly, German wind capacity has mushroomed, from 56 MW in 1990 to more than 14,600 MW in 2003.[8] (See Figure 6, page 30.) Germany passed the United States to become the world's leading wind energy producer in 1997. Wind power now meets more than 6 percent of Germany's total electricity demand, up from 3 percent in late 2001.[9] In the northern reaches of the country, where most of the development is concentrated, wind power provides as much as 29 percent of annual electricity needs, close to nuclear power's share for Germany as a whole.[10] As for PV, since 1992 it has grown at an average annual rate of nearly 47 percent. Germany ended

FIGURE 6

Wind Power Capacity Additions in Germany, Spain, and the United States, 1980–2003

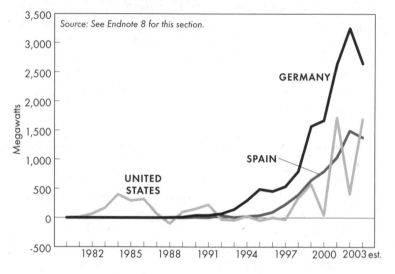

Source: See Endnote 8 for this section.

2003 with 417 MW of PV capacity, mostly on-grid, and is now second only to Japan in PV capacity.[11] To meet rapidly rising demand, major German manufacturers plan to expand PV manufacturing facilities significantly over the next few years, which will further reduce costs and increase employment.[12]

In 2002 alone, the sales in German renewable energy industries totaled nearly $11 billion.[13] Some 45,000 people worked in Germany's wind industry by early 2003, one-fifth of them hired the previous year.[14] The 100,000 Roofs program alone created an estimated 10,000 new jobs, at a cost of $27,000 per position, and Germany accounts for most of Europe's PV installations.[15] Germany also boasts Europe's largest shares of biogas capacity and solar thermal water heaters.[16] With so many Germans employed in renewables industries or owning shares in wind turbines, solar, or other projects, renewable energy enjoys broad support.

Germany has pledged to reduce its CO_2 emissions 21 percent below 1990 levels by 2010, and plans to accomplish much of this through increased use of renewable energy.[17] To date, renewable energy is responsible for 50 million tons (6.25 percent) of Germany's total CO_2 emissions reductions.[18] The government aims for wind power to meet 25 percent of national electricity needs by 2025, with a target of 25,000 MW of wind capacity offshore, and also considers solar PV as a viable long-term option for large-scale power generation.[19] By 2050, Germany intends to meet at least half of its total energy needs with renewable sources.[20] The total costs of market development programs for all new energy technologies through 2050 appear to be significantly lower than the total spent over all years on coal.[21]

Japan's story with PV is similar to Germany's experience with renewables. It rose from a minor player in the early 1990s, manufacturing PV units primarily for use in calculators and watches, to become the world's largest producer and user in less than a decade. With far less land area and about half the solar insolation of California, Japan now has three times as much PV capacity as the entire United States.[22]

Driven by concerns about energy security and climate change, Japan has enacted effective and consistent policies to promote PV, and has retained them even through major budget crises. The "New Sunshine" program was established in 1992 to introduce renewable energy throughout the country. Targets were set and a new net metering law enacted to require utilities to purchase excess PV power at the retail rate.[23] Two years later, Japan launched the "Solar Roofs" program to promote PV through low-interest loans, a comprehensive education and awareness program, and rebates for grid-connected residential systems in return for data about systems operations and output. At the time, Japan had about 31 MW of installed PV and accounted for less than one-fourth of global PV manufacture.[24]

The residential rebates started at 50 percent of installed costs and declined gradually over time. In 1997, rebates were extended to owners and developers of housing complexes as well, and Japan became the world's largest supporter of PV with a seven-fold increase in funding for the expanded program, which became known as the "70,000 Roofs" program. The budget for the residential PV dissemination program increased from $20 million in 1994 to a peak of $219 million in fiscal year 2001; the FY 2004 budget is just under $49 million.[25]

Government promotion of PV has included publicity on television and in print media.[26] The national government has also encouraged the use of PV in government office buildings, and many local governments provide PV subsidies and low-interest loans.[27] As a result of Japan's net metering law, between April 2001 and March 2002 alone, Japanese electric power companies bought more than 124 GWh of surplus PV power at retail rates.[28]

The goal of Japan's PV program has been to create market awareness and stimulate production in order to reduce costs through economies of scale and technology improvements, and thereby enable large-scale power generation and the export of PV products to the rest of the world. Japan is now the world's leader in the manufacture and use (i.e., capacity) of solar PV, having surpassed the United States in both respects in the late 1990s.[29] (See Figure 7 regarding capacity.)

A number of policies have contributed to PV's success in Japan, but the 70,000 Roofs program is considered by some to be the most important government PV program in history.[30] While some subsidies remain at the national, state, and municipal levels, the Solar Roofs program ended officially in 2002 after exceeding all objectives.[31] The program resulted in the installation of more than 144,000 residential systems, with capacity totalling 424 MW.[32] Nearly 43,000 households applied for program funding in 2002 alone, when subsidies were down to about $1/W.[33] Primarily due to the residential program, total installed PV capacity in Japan has increased an average of more than 43 percent annually since 1993, totalling 887 MW by the end of 2003.[34] The government aims for total PV installations to reach 4,820 MW by 2010.[35]

Despite the decline in subsidies, new home installations continue to rise as costs fall, and Japan's PV market is expected to continue growing by 20 percent annually over the next several years.[36] By some accounts, small-system costs in Japan have dropped more than 80 percent since 1993, far more rapidly than the decrease in global module costs over this period.[37] Installed

FIGURE 7

Photovoltaic Capacity Additions in Japan, Germany, and the United States, 1993–2003

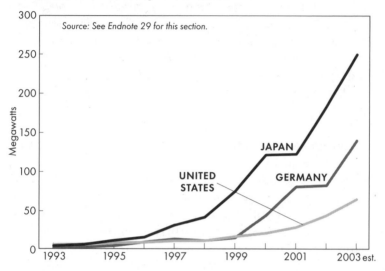

Source: See Endnote 29 for this section.

costs of residential grid-connected systems have fallen from $11/W in 1995 to about $5.50/W in 2003, not including subsidies.[38] As a result, on-grid PV-generated power in Japan, at 11–15 cents/kWh, is now cheaper than retail electricity.[39]

To keep up with rising demand, Japanese PV manufacturers have invested significantly in plants and equipment, increasing their production capacity by nearly 47 percent in 2002 and 45 percent in 2003.[40] Japan was responsible for half of global PV production in 2003, and Sharp has been the world's leading producer of solar cells since 2000.[41]

Policy Lessons From Around the World

It is difficult to claim that something is impossible once it has already occurred. This is why it is globally significant that two of the world's largest economies transformed themselves from laggards to leaders in renewable technologies in less than a decade. What Germany and Japan have accomplished can be replicated elsewhere—with the right mix of policies.

For renewable energy to make as large as possible a contribution to economic development, job creation, lower oil dependence, and reduced greenhouse gas emissions, it is essential to improve the efficiency of technologies, reduce their costs, and develop mature, self-sustaining industries. Today's energy markets frustrate efforts to achieve these goals because of lack of access to the electric grid at reasonable prices, high initial costs compared to conventional energy sources, and widespread ignorance about the scale of resources available, the pace of development of renewable technologies, or the potential economic advantages of renewable energy.

The dramatic successes seen in Germany and Japan stem from a range of policies introduced to address these barriers. They demonstrate that policies play a far greater role than a nation's resource base in determining its renewable energy generation. They also demonstrate that, in addition to the global learning curve that is driving down technology costs, there is a national learning curve that drives domestic costs down even faster and further as countries develop domestic industries to manufacture, install, and maintain renewable systems using local equipment and labor. The experiences of Germany, Japan, and other countries provide an array of promising policy options that can be disseminated around the world.

There are five major categories of relevant policies:

1. Market Access and Obligations. As Germany's experience demonstrates, access to the market is imperative for renewables to gain a foothold. Two main types of regulatory policies have been used to open the electric grid to renewables. The first is pricing laws, which guarantee renewable producers fixed, minimum prices and obligate electric utilities to provide grid access to renewable energy plants. Fixed payments, or tariffs, are paid over several years, and often decline over time to reflect cost reductions. The costs of the pricing system are covered by energy taxes or an additional per-kilowatthour charge on electricity consumers.

Spain, Denmark, and several other countries have enacted similar pricing laws. Before Spain passed a pricing law in 1994, few wind turbines were spinning in the Spanish plains and mountains, but by the end of 2000 the country ranked third in the world in wind installations.[1] Spain generated 5 percent of its electricity from wind in 2003 and is now home to some of the world's largest turbine manufacturers.[2] Denmark now generates more than 20 percent of its electricity from wind and has long been the world's wind-turbine leader.[3]

The marriage of a guaranteed market and long-term minimum payments has reduced investment risks, making it profitable to invest in renewable technologies and easier to obtain financing. By creating demand for renewable electricity, the pricing law has attracted private investment for R&D, spread the costs of technology advancement and diffusion relatively evenly across populations, and enabled the production scale-ups and the installation, operation, and maintenance experience needed to bring down the costs of renewable technologies and generation.

The second type of regulatory access policy, the quota system, works in reverse of pricing laws: governments set targets and let the market determine prices. Typically, governments mandate a minimum share of capacity or generation to come from renewable sources. As with pricing systems, the additional costs of renewable energy are borne by taxpayers or electricity consumers.

With the most common form of quota system (such as the Renewables Portfolio Standard, or RPS, used in several U.S. states), investors and generators comply with the quota by installing capacity, purchasing renewable electricity through a bidding process, or buying "green certificates" or "renewable energy credits." Generally certificates are awarded to producers for the renewable electricity they generate, and add flexibility by enabling utilities and customers to trade, sell, or buy credits to meet obligations. They can add value to renewable installations by creating a paper market separate from electricity sales, and can allow for trading and expanding renewable energy markets between states or countries.

Texas's RPS is primarily responsible for wind energy's rapid growth there since 1999, when the state required that 2,000 MW of additional renewable capacity be installed within a decade. Texas was more than halfway there with wind alone by mid-2002, and the target will likely be met long before 2009.[4] But the mandates alone have done little to encourage the use of more expensive technologies such as PV, despite vast solar resources in Texas. About one-fourth of U.S. states have enacted RPS laws, many of them experiencing less success than Texas.[5] RPS systems with enforced penalties for noncompliance and specific technology quotas are most effective at ensuring that targets are met and a range of technologies is developed. Mandated quotas are now used in several other countries as well, including Japan, the United Kingdom, Italy, and Australia.

Under tendering systems, another type of quota system, potential project developers bid to a public authority for contracts to fulfill their government mandate. Projects that are considered viable and that compete successfully on price terms against other bidders are offered contracts to receive a guaranteed price per unit of electricity generated. The government often covers the difference between the market reference price and the winning bid, and contracts are generally awarded for a period of several years.

The United Kingdom enacted a tendering system in 1989, and between 1990 and 1998 renewable developers competed for contracts in a series of bidding rounds. While this system made it easier to obtain financing, it created flurries of activity followed by long lulls with no development, making it difficult to build a domestic turbine manufacturing industry and infeasible for small firms or cooperatives to take part. Competition to reduce costs and win contracts led developers to seek sites with the highest wind speeds, often areas of scenic beauty, which increased public opposition to wind energy and made it harder to obtain project permits. The lack of deadlines allowed winning contractors to wait years for costs to fall before building projects.[6] When the program ended in 1999, more than 2,670 MW of wind capacity were under contract, but only 344 MW had been installed.[7] A new quota

system based on renewable certificates has since been established, and capacity reached 649 MW by the end of 2003.[8] This
system is expected to provide a significant boost for wind
power, but it is still too early to tell what the impacts will be.

A variation on pricing laws, net metering, can be used in
conjunction with quota systems. Net metering permits consumers to install small renewable systems and sell excess electricity into the grid at wholesale market prices. It differs from
pricing laws primarily in scale and implementation, and is
available in several countries, including Japan, Thailand, and
Canada. At least 38 U.S. states, including California and Texas,
have enacted such laws.[9] Success in attracting new renewable
energy investments and capacity depends on the limits set on
participation (capacity caps, number of customers, or share of
peak demand); the price paid, if any, for net excess generation;
the existence of grid-connection standards; enforcement mechanisms; and other available incentives.

Of these regulatory options, pricing laws have consistently
proved to be the most successful to date. Although they have
not succeeded in every country that has enacted them, those
countries with the most significant growth and the strongest
domestic industries have had pricing laws. While at least 48
countries have installed wind capacity, just three—Germany,
Denmark, and Spain, all with pricing laws—account for more
than 84 percent of total wind capacity installed in the European Union, and 59 percent of global capacity.[10]

Pricing laws can be designed to account for changes in
the marketplace, encourage steady growth of small- and
medium-scale producers, encourage private sector investment
in R&D, offer ease of financing, and enable even average citizens to benefit from investments in renewable energy projects.
Although some argue that pricing systems are more costly
than quota systems, costs depend more on policy details than
system type, and several studies have concluded that the average additional costs per household of the German pricing law
have been minimal.[11] Further, both system types involve subsidies as they create protected markets for renewables. Quota
systems have not been in use as long, so there is a lack of

experience with them, but to date the record of such systems is more uneven and reveals a tendency toward boom-and-bust markets. However, success is determined by system details and by other policy mechanisms enacted in parallel.

2. Financial Incentives. Market compensation mechanisms (tax credits, rebates, loans, or payments) that subsidize investment in a technology or the production of power have been used extensively in Europe, Japan, the United States, and India. To encourage investment in renewables in the 1980s, the U.S. government and California offered investors credit against their income taxes, allowing them to recoup a significant share of their money in the first few years and reducing their risk. The credits played a major role in a wind boom that many called California's second gold rush. The lessons learned and economies of scale gained through this experience advanced wind technology and reduced its costs.[12] But enormous tax breaks and a lack of technology standards encouraged fraud and the use of untested and substandard equipment, some of which never generated a kilowatthour of electricity.[13]

India saw a similar boom a decade later, sparked by a combination of investment tax credits, financing assistance, and accelerated depreciation.[14] India is now the world's fifth-largest producer of wind power and has developed a domestic manufacturing industry. As in California, however, investment-based subsidies and a lack of standards or production requirements led wealthy investors to use wind farms as tax shelters, and several projects performed poorly despite the significant technology advancements since the early 1980s.[15]

Some countries, like Japan, have subsidized investment through rebates and have seen dramatic successes, with PV in particular. Twenty-four U.S. states offer PV rebates that cover a large share of the costs—up to 50 percent in California and Massachusetts, and 70 percent in New Jersey and New York.[16] Due to rebates and resulting cost reductions, some California builders now include PV on homes in entire subdivisions.[17] (See Sidebar 3.)

Since 1994, the U.S. government has offered a production tax credit to those who supply wind-generated electricity

SIDEBAR 3

Public Benefit Funds and Bond Initiatives

Fifteen U.S. states have established accounts to finance renewable energy projects, funded through small per-kilowatthour surcharges on electricity consumption. In early 2004, 15 such public benefit funds from 12 states announced formation of the Clean Energy States Alliance; CESA will invest $3.5 billion over the next decade to create larger markets than each fund could promote individually, thereby spurring innovation and producing more jobs.

California has the largest fund, created in 1996 as part of the state's electricity restructuring legislation. It has enabled California to provide production payments for existing and new renewable energy projects, as well as rebates for consumers who buy certified green power and for people investing in "emerging renewables" (PV, small-scale wind, solar thermal electric, and fuel cells powered with renewable hydrogen). Since 1998, California's PV program has helped reduce system costs by 50 percent, has dramatically increased grid-connected capacity, and has increased the number of in-state PV manufacturing, distribution, and installation companies. California is now the third-largest PV market in the world, with 10 times more installed PV capacity than any other U.S. state.

Another California program driving PV is the San Francisco Solar Bond Initiative. In 2001, city voters overwhelmingly approved a $100-million bond program to purchase renewable energy for public facilities. A combination of bulk purchasing and bundling of PV with wind energy and efficiency measures means that energy savings will cover the additional costs associated with PV. The program aims to increase public awareness, create jobs, drive down PV costs through economies of scale, and to make the city a world leader in the use of clean energy. Several other U.S. cities and states are considering following San Francisco's lead.

Sources: See Endnote 17 for this section.

to the grid. The credit has encouraged wind development, but only in states with additional incentives, and it provides greater benefit to producers with higher income levels and tax loads.[18] California has enacted a per-kilowatthour production payment, rather than a tax credit, for some existing and new renewable projects. It is financed through a small charge on electricity use, meaning that Californians share the cost of the program according to their consumption level. Provided that such payments are high enough to cover the costs of renewable generation and are guaranteed over a sufficient period, such a policy integrates several key elements of a pricing law and may be similar in effect (and perhaps more politically feasible in some countries).[19]

Experience to date demonstrates that payments and rebates are preferable to tax credits. Unlike tax credits, the benefits of payments and rebates are equal for people of all income levels. Further, production incentives are generally preferable to investment subsidies because they promote the desired outcome, energy generation. However, policies must be tailored to particular technologies and stages of maturation, and investment subsidies can be helpful when a technology is still relatively expensive, as with PV in Japan.[20] All subsidies should be gradually reduced and phased out to encourage cost reductions.

Financing assistance in the form of low-interest, long-term loans and loan guarantees is also essential to address the high up-front capital costs of renewables. Lowering the cost of capital can reduce the average cost of electricity and the risk of investment, as seen in Germany. Even in the developing world, all but the very poorest people are able and willing to pay for reliable energy services, but they need access to low-cost capital. According to PV companies in South Africa, Indonesia, India, and the Dominican Republic, up to 50 percent of prospective purchasers can afford systems if reasonable third-party financing is available; otherwise, only 2–5 percent can buy them.[21] Thus the availability of financing could increase PV use in some countries by 10-fold or more, and the impacts could be similar with other renewables.

One of the most successful means for disseminating household-scale renewable technologies in rural China has been local public-private bodies that offer technical support, materials sales, subsidies, and government loans for locally manufactured technology. They frequently provide revolving credit, with repayment linked to the timing of a household's income stream.[22] In India, the terms of long-term, low-interest loans vary by technology, with the most favorable for PV. In addition, the national government has obtained bilateral and multilateral funding for large-scale projects, particularly wind.[23]

3. Education and Information Dissemination. Even if governments offer generous incentives and low-cost capital, people will not invest in renewable energy if they are uninformed—or misinformed—about resource availability, technology development, the advantages and potential of renewables, the fuel mix of the energy they use, and the incentives themselves. During the 1980s, several U.S. states offered substantial subsidies for wind energy, including a 100-percent tax credit in Arkansas, a state with enough wind to generate half of its electricity. But these subsidies evoked little interest due to ignorance about wind resources.[24] By contrast, it was wind resource studies in California, Hawaii, and Minnesota that generated interest in wind energy there.[25] Cloudy Germany has more solar water heaters than sunnier Spain and France, mostly because public awareness of the technology is so much higher in Germany.[26]

Inexperience (or bad experiences) have left many with a perception that renewables do not work, are inadequate to meet their needs, are too expensive, or are too risky as investments. Above all, it is essential that government leaders recognize the inherent value of renewable energy. Then governments, non-governmental organizations, and industry must work together to educate labor organizations about employment benefits, architects and city planners about ways to incorporate renewables into building projects and their value to local communities, agricultural communities about their potential to increase farming incomes, and so on.

Training and certifying workers are also essential, to ensure that people are available to manufacture, install, and maintain renewable energy systems. Austrian students learn about renewable energy in schools and universities, and many German vocational schools have renewable energy programs.[27] The Indian government has also established training programs, and has used print, radio, songs, and theater to educate the public about the benefits of renewable energy and

government incentives. The Solar Finance Capacity Building Initiative educates Indian bank officials about solar technologies and encourages them to invest in projects.[28]

The problems thwarting renewables (and their solutions)
are not necessarily unique to particular countries or settings.[29]
Thus it is essential to share information at all levels regarding
technology performance and cost, capacity and generation statistics, impacts of renewable energy on society, and policy
successes and failures in order to increase awareness and avoid
reinventing the wheel each time. While several countries do
this on a national level, a centralized global clearinghouse for
such information is clearly needed.

4. Stakeholder Involvement. Public participation in
policymaking, project development, and ownership also
increases the odds of success. In Germany and Denmark,
where individuals (singly or as members of cooperatives) still
own many of the turbines installed, there is strong and broad
public support for wind energy. As of 2002, about 85 percent
of the installed wind capacity in Denmark was owned by
farmers or cooperatives, and at least 340,000 Germans had col-
lectively invested nearly $14 billion in renewable energy proj-
ects.[30] Through cooperatives, people share in the risks and
benefits of renewable energy; often avoid problems associated with obtaining financing and paying interest; play a
direct role in the siting, planning, and operation of equipment;
and gain a sense of pride and community.[31]

Public participation and a sense of ownership are as
important in the South as in the North. When technologies
are forced on people without consultation regarding their
needs or are donated as part of an aid package, people often
place little value on them and feel they have no stake in
maintaining them. But when individuals and communities play
a role in decision making and ownership, they are literally
empowered and become invested in the success of the technologies. Local participation in and ownership of solar mini-
grid projects in Nepal and the Indian islands of the Sundarbans
have helped ensure the projects' success and have eliminated
electricity theft.[32]

5. Industry Standards, Permitting, and Building Codes. Standards are essential to prevent inferior technologies from entering the marketplace and generate greater confidence in products, thereby reducing risks and attracting investors. Technology standards for wind turbines, for example, can apply to everything from turbine blades, electronics, and safety systems to performance and compatibility with the transmissions system. Largely due to pressure from the wind industry, Denmark adopted wind turbine standards in 1979. These standards are credited with playing a major role in Denmark's becoming the world's leading turbine manufacturer.[33] Germany established an investment tax credit for wind energy in 1991, and turbine standards and certification requirements prevented the quality control problems experienced in California and India. Eventually, technology standards for all renewable technologies should be established at the international level.

Standards and planning requirements can reduce opposition to renewables if they address potential concerns such as noise and visual or environmental impacts. Such laws can be used to reserve specific locations for development or to restrict areas at higher risk of environmental damage or injury to birds, for instance, reducing uncertainty about project siting and speeding the planning process.[34] The United Kingdom provides the best example of how the lack of planning regulations can paralyze an industry: despite the best wind resources in Europe, the nation added little wind capacity under its early quota system, largely because a lack of planning regulations virtually halted the permitting process.[35]

Building codes and standards can also be designed to require renewables' incorporation into building designs and planning processes. London Mayor Ken Livingstone spearheaded a proposed strategy that, if enacted, will require major developments there to incorporate solar energy or be designed for easy future installation.[36] Including wiring and other hardware in new buildings to make them solar-ready adds little to construction costs while making it easier and far cheaper to install such systems later. And efficiency standards can facilitate the use of renewable energy by making the scale more manageable (so renewables can more easily satisfy energy needs), and by reducing the load so that it is easier to bear higher costs per unit of output.

Changing Government Approaches to Energy Policy

Energy markets are not now and never have been fully competitive and open. Discriminating standards, regulations, government purchases, past investments in infrastructure and long-term subsidies for conventional energy, and the failure to internalize external costs and benefits all act as obstacles to the advancement of renewable energy. Thus, perhaps the most important step governments can take to advance renewables is to transform their perspectives and approaches to energy policy. Governments must eliminate inappropriate, inconsistent, and inadequate policies that favor conventional fuels and technologies and that fail to recognize the social, environmental, and economic advantages of renewable energy.

One of the most important steps governments can take to level the playing field is to eliminate subsidies for conventional energy. Mature technologies and fuels should not require subsidization, and every dollar spent on conventional energy is a dollar not invested in clean, secure, and sustainable renewable energy. In the mid-1990s, governments worldwide were handing out $250–300 billion annually to subsidize fossil fuels and nuclear power.[1] Since then, several transitioning and developing countries have reduced energy subsidies significantly, but global subsidies for conventional energy remain many magnitudes higher than those for renewable energy.[2]

At the international level, the Global Environment Facility allocated $650 million to renewable energy projects in developing countries between 1992 and 2002.[3] This is a small fraction of global investments in carbon-intensive energy projects through international financial institutions like the World Bank and taxpayer-funded export credit agencies. Over the past decade, World Bank funding for fossil fuel projects (totalling $26.5 billion) has exceeded that for renewable energy and efficiency by a factor of 18.[4]

In most cases, it is less a matter of finding new money to invest in renewable energy than of transferring money flows from conventional energy to renewables. Each year, an estimated $200–250 billion are invested in energy-related infrastructure to replace existing capital stock and meet ever-rising demand, and another $1.5 trillion is spent on energy consumption; nearly all of this goes to conventional energy.[5] Hundreds of millions of people in the developing world spend about $20 billion every year on makeshift solutions such as candles, kerosene lamps, and batteries.[6] The International Energy Agency projects that $16 trillion will be invested worldwide in energy-supply infrastructure between 2001 and 2030.[7] Even small shifts in these expenditures would have a tremendous impact on renewable energy markets and industries.

Next, pricing structures must account for the significant external costs of conventional energy and the advantages of renewable energy. Germany has begun to do this through its pricing law and other countries do so with energy or carbon taxes. And as the single largest consumers of energy in most or all countries, governments should purchase ever-larger shares of energy from renewables and thereby set an example, increase public awareness, reduce perceived risks associated with renewable technologies, and reduce costs through economies of scale.

Finally, policies designed to advance renewable energy can fail if they are not well formulated or are inconsistent, piecemeal, or unsustained. For example, because early investment credits in California were short-lived and extensions were often uncertain, many equipment manufacturers could not begin mass production for fear that credits would end too soon.[8] When the incentives expired, interest waned and the industries and markets died with them. The U.S. Production Tax Credit for wind energy has expired several times, only to be extended months later. As a result, the credit has stimulated wind capacity growth but has created cycles of boom and bust in the market, with busts causing suspension of projects, worker layoffs, and loss of momentum in the industry.

This on-and-off approach to renewables has made the development of a strong U.S. industry a challenge, at best. In India, uncoordinated and inconsistent state policies, and bottlenecks imposed by state electricity boards, have impeded renewable energy development.[9] Even in Denmark, years of steady wind-energy growth ended in 1999 when the government changed course and doubt overtook years of investor confidence. The future of some planned offshore wind farms is now uncertain, as is Denmark's target to produce half its electricity with wind by 2030. The number of jobs in the domestic industry will probably decline over the next few years.[10] These changes are due not to the technologies themselves but to inconsistencies and failures in policy.

Consistent policy environments are necessary for the health of all industries. Consistency is critical for ensuring continuous market growth, enabling the development of a domestic manufacturing industry, reducing the risk of investing in a technology, and making it easier to obtain financing. It is also cheaper, because higher incentives might be required to coax investors back into the market as uncertainty increases the perception of risk, and because stop-and-go policies force funds to be reappropriated, new programs administered, information distributed to stakeholders, and so on.[11] Government commitment to developing renewable energy markets and industries must be strong and long-term, just as it has been with fossil fuels and nuclear power. (See Sidebar 4.)

Unlocking Our Energy Future

In early January 2004, the U.S. unmanned rover *Spirit* touched down on the surface of Mars and within days began relaying to Earth dramatic photographs of a red, rock-strewn surface, distant hills, and a rust-colored sky from 170 million kilometers (106 million miles) away.[1] Humanity has clearly

---

SIDEBAR 4

Forging a New Energy Future

- Enact renewable energy policies that are consistent, long-term, and flexible, with enough lead time to allow industries and markets to adjust.

- Emphasize renewable energy market creation.

- Provide ready access to the electric grid at prices that reflect full costs of conventional energy and supply sufficient incentive to stimulate renewable energy market growth.

- Provide financing assistance to reduce up-front costs through long-term, low-interest loans, through production payments for more advanced technologies, and through investment rebates for more expensive technologies such as PV, with gradual phaseout.

- Disseminate information regarding resource availability, the benefits and potential of renewable energy, capacity and generation statistics, government incentives, and policy successes and failures at local, national, and international levels.

- Encourage individual and cooperative ownership of renewable energy projects and ensure that all stakeholders are involved in the decisionmaking process.

- Establish standards for performance, safety, siting, and buildings.

- Incorporate all costs into the price of energy and shift government subsidies and purchases from conventional to renewable energies.

---

established a presence on two planets—and one of them is powered primarily by renewable energy. PV modules enable *Spirit* and its twin, *Opportunity*, to roll across the planet's surface, operate sophisticated cameras and rock abrasion tools, analyze materials, and send valuable data and photographs back to Earth. In fact, without energy from the sun and high-tech, reliable, renewable technologies such as PV, space exploration itself would be impossible.

It will be a long time before renewables achieve the penetration level on Earth that they currently enjoy on Mars, but renewable energy is coming of age even on our planet. After more than a decade of double-digit growth, renewable energy is a multibillion-dollar global business. Wind power is leading the way in many nations, supplying more than 20 percent of the electricity needs in some regions and countries. It represents almost half of global investment in renewable technologies, and is now cost-competitive with conventional energy technologies. Solar cells are already the most affordable option for getting modern energy services to hundreds of millions of people in developing countries and are competitive on-grid in Japan today. Their costs continue to fall rapidly.

Renewable technologies are attracting the funds of venture capitalists and multinational corporations alike. The major oil companies BP and Royal Dutch/Shell have invested hundreds of millions of dollars in renewable energy development. While this is a fraction of what they devote to oil and gas, it is a move in the right direction. General Electric has also become a large player, supplying 15 percent of the global wind turbine market in 2003, and is beginning to enter the PV market.[2] In early 2004, the largest U.S. financial institution, Citigroup, announced plans to begin investing in renewable energy.[3] Worldwide, investment in new renewable energy technologies is expected to increase more than four-fold between 2003 and 2012, to $85 billion annually.[4]

Whether renewable energy capacity and investment continue to grow at current levels will hinge largely on policy decisions by governments around the world. Expansion during the past decade has occurred because of substantial policy changes in a half-dozen countries, and those nations alone are not large enough to sustain the growth required to propel renewables into the mainstream worldwide. But recent developments suggest that political support for renewables is rising around the world.

One example is Europe, the engine of growth for the global wind industry. In the United Kingdom, which until recently was a European straggler on renewables, Prime Minister Tony Blair has called his nation's investment in renewable energy "a major down-payment in our future" that will "open up huge commercial opportunities."[5] The European Union aims for renewables to generate 22 percent of Europe's electricity by 2010.[6] Elsewhere, China has upped its wind energy targets and

TABLE 3

Renewable Energy Targets and Recent Totals in Selected Countries/Regions

Country/Region	Targets for Renewable Energy	Recent Totals
California, U.S.	20% electricity from new renewables by 2017	12% (2002)
China	4,000 MW wind by 2010; 20,000 MW wind by 2020	568 MW (2003)
European Union	22.1% electricity by 2010; 12% total energy by 2010	14% electricity (1999); 6% energy (1997)
Germany	20% electricity by 2020; 50% total energy by 2050	6.8% electricity (2002)
Japan	4,830 MW of PV by 2010	887 MW (2003)
Latin America, Caribbean	10% total energy from new renewables by 2010	NA
Navarra, Spain	97% electricity by 2005	55% (2002)
Thailand	21.2% total energy by 2011	19.8% (2001)

Notes: Values are for all types of renewables unless otherwise noted. For California, RPS mandate for investor-owned utilities only; credit for existing but not new small hydropower plants. For Latin America and Caribbean countries above target must maintain their current share. Source: See Endnote 9 for this section.

plans to invest $1.2 billion in PV over the next five years.[7] India has proposed that 10 percent of annual additions to electric capacity come from renewables by 2012.[8] In Latin America, Brazil is leading the way with a comprehensive and ambitious renewable energy law.[9] (See Table 3.)

Even in the United States, despite an oil-oriented White House, nearly half the members of Congress have joined the Renewable Energy and Energy Efficiency Caucus.[10] Although this political support has not yet translated into the needed federal legislation, many states—including Arizona, California, Nevada, New York, and Texas—have enacted pioneering laws, and more and more governors are professing the benefits of renewable energy for their states, from energy security and jobs to reduced dependence on imported oil.[11]

Despite the substantial strides being made in technology, investment, and policy, renewables continue to face a credibility gap. Many people remain unconvinced that renewable energy can one day be harnessed on a scale that would meet most of the world's energy needs. Renewable energy sources appear too ephemeral and sparsely distributed to provide the energy required by a modern post-industrial economy. But those assumptions are outdated. In the words of Paul Appleby, formerly with BP's solar division, "the natural flows of energy are so large relative to human needs for energy services that renewable energy sources have the technical potential to meet those needs indefinitely."[12]

The G8 Renewable Energy Task Force projects that in the next decade up to a billion people could be served with renewable energy.[13] BP and Shell have predicted that renewable sources could account for 33 to 50 percent of world energy production by 2050, with stable regulatory frameworks.[14] And David Jones of Shell has forecast that renewables could emulate the rise of oil a century ago, when it surpassed coal and wood as the primary source of energy.[15]

Not only is renewable energy alone sufficient to meet all of today's energy needs thousands of times over, harnessing it is not particularly land- or resource-intensive. Theoretically, all U.S. electricity could be provided by wind turbines in Kansas, North Dakota, and South Dakota, or with solar energy on a plot of land 100 miles square in Nevada.[16] Farming under the wind turbines could continue as before, while farmers enjoyed the supplementary revenues from spinning wind into electricity. In cities around the world, much of the local power needs could be met by covering existing roofs with solar cells—requiring no land at all. Additional energy will be provided by wind and ocean energy installations located several kilometers offshore, where the energy flows are abundant.

The other credibility gap that must be bridged is how to provide renewable energy when and where it is needed. How do you get wind or sunshine into a gas tank, for example, and on a still, dark night? That question may have been answered by automobile and energy companies around the world. Just as electricity enables us to use and transport renewable energy today, hydrogen offers a promising option—once costs have dropped significantly and infrastructure is in place—for producing fuel from renewable energy, storing it underground, and carrying it by pipeline to cities and factories. Major automobile manufacturers are developing hydrogen internal combustion engines and fuel cell-powered cars that will emit only water from their tailpipes. DaimlerChrysler, Honda, Toyota, and GM now expect to have their first commercial fuel-cell cars available by 2010.[17]

The next challenge for renewables will be how to enter the mainstream and overtake fossil fuels in light of investments already made in conventional infrastructure that will be operable for decades to come. But infrastructure and power capacity are being replaced or added continuously, and this is where a significant shift toward renewable energy must begin in the developing and industrial worlds alike. A recent study determined that renewables could supply 20 percent of Europe's energy demand and 33 percent of its electricity by 2020. To meet the EU's targets for 2010 and proposed goals for 2020, 52.5 percent of new power capacity installed from 2001 through 2010 and 61 percent installed from 2011 through 2020 would need to be renewable. It is estimated that avoided fuel and environmental costs could equal the projected costs of investment.[18] Another study concluded that Europe could phase out nuclear power and reduce carbon emissions 80 percent by 2050 through a transition to renewable energy that, if external costs were incorporated, would be far cheaper than continuing with business as usual and would provide new jobs as well. However, this transition will be possible only if the necessary steps down this road are begun as soon as possible.[19]

In early 2001, the Intergovernmental Panel on Climate Change released its most recent report, confirming that in order to stabilize the world's climate, "eventually CO_2 emissions would need to decline to a very small fraction of current emissions"—meaning close to zero.[20] If the world is to achieve this goal—which it must—countries must begin today, not tomorrow, to make the transition to a renewable, sustainable energy future. We have a brief window of opportunity to start down the path to a more sustainable world—one in which rising demand for energy is met without sacrificing the needs of current and future generations and the natural environment.

We still have a long way to go to achieve this vision. Today most of the world is locked into a carbon-based energy system that is neither better nor necessarily cheaper than renewable energy, but merely the legacy of past policies and investment decisions. Breaking with this past will not be easy. But Germany, Japan, and other countries are proving that change is indeed possible and that it can happen rapidly. The key is ambitious, forward-looking, consistent government policies that drive demand for renewable energy, create a self-reinforcing market, and propel renewables into the energy mainstream during the 21st century.

Endnotes

Introduction

1. $20.3 billion is Worldwatch estimate based on wind and PV capacity additions in 2003, an assumed 3-percent growth rate for other renewables, and on 2002 renewables estimates from Eric Martinot, "Global Renewable Energy Markets and Policies," forthcoming in *New Academy Review*, Spring 2004. The estimate was developed using Eric Martinot's methodology, described at www.martinot.info/markets.html, viewed 29 February 2004, and in an e-mail from Martinot to author, 29 February 2004. One-sixth share of total investment is calculated by Worldwatch based on the above and on the assumption in Martinot (ibid.) that total global investment in conventional power generation is $100–150 billion annually.

2. Two billion from José Goldemberg, "Rural Energy in Developing Countries," in José Goldemberg, ed., *World Energy Assessment: Energy and the Challenge of Sustainability* (New York: United Nations Development Programme, 2000), p. 348.

The Approaching Train Wreck—and How To Avoid It

1. "Energy-Hungry China Braces for Power Struggle as Winter Draws Near," *China Daily*, 8 December 2003; and Peter S. Goodman, "China's Dark Day Days and Darker Nights: Industrial Growth Exceeds Supply of Electrical Power," *Washington Post*, 5 January 2004.

2. Feng Jianhua, "Energy Crisis? The simultaneous shortage of oil, electricity, and coal is an indicator that China needs to improve its energy security," *Beijing Review*, March 2004, at www.bjreview.com.cn/200403/Business-200403(A).htm, viewed 5 March 2004; slowing economic growth from Goodman, op. cit. note 1.

3. World Bank, *Clear Water, Blue Skies: China's Environment in the New Century*, China 2020 Series (Washington, D.C.: 1997).

4. Daniel Yergin and Michael Stoppard, "The Next Prize," *Foreign Affairs*, November/December 2003, p. 109. China's power demand is expected to rise 15 percent in 2004 and 11 percent in 2005; Kimberly Song and Kathy Chen, "A Power Shortage Starts To Hinder China's Output," *Wall Street Journal*, 9 December 2003.

5. Electricity increase from Goodman, op. cit. note 1; oil increase from John C.K. Daly, "UPI Energy Watch," United Press International, 26 February 2004, at http://washingtontimes.com/upi-breaking/20040226-040855-1015r.htm, viewed 17 March 2004.

6. Goodman, op. cit. note 1.

7. Manoj Kumar, "Tryst With Developing World Consumers: A Case Study of India," *The ICFAI Journal of Marketing Management*, November 2002; and Victor Mallet, "China Unable To Quench Thirst for Oil," *Financial Times*, 20 January 2004.

8. International Energy Agency, *World Energy Outlook 2002* (Paris: 2002).

9. Oil peaking from "Analysts Claim Early Peak in World Oil Demand," *Oil & Gas Journal Online*, 12 August 2002; and Bruce Stanley, "Oil Supply Seen Set To Fall," *Washington Times*, 28 May 2002; remaining resources expensive to extract from Charles Hall et al., "Hydrocarbons and the Evolution of Human Culture," *Nature*, vol. 426, 20 November 2003, p. 322.

10. Shimp quoted in "Feature—Solar Power To Challenge Dominance of Fossil Fuels," *Reuters*, 9 August 2002.

11. Greater threat than terrorism from David Ljunggren, "Global Warming Bigger Threat Than Terrorism," *Reuters*, 6 February 2004; and Steve Connor, "U.S. Climate Policy Bigger Threat to World Than Terrorism," *The Independent*, 9 January 2004; most important global challenge is United Nations statement regarding ministers attending international climate negotiations in Milan, Italy, December 2003. United Nations Framework Convention on Climate Change press release (Milan: 12 December 2003), cited in *Wind Directions*, January/February 2004, p. 41.

12. Intergovernmental Panel on Climate Change, *Climate Change 2001: The Scientific Basis* (Cambridge, U.K.: Cambridge University Press, 2001), pp. 223–34.

13. Losses due to natural disasters from UN Environment Programme (UNEP), "Financial Sector, Governments, and Business Must Act on Climate Change or Face the Consequences," press release (Nairobi: 8 October 2002); and UNEP, "Weather Related Natural Disasters in 2003 Cost the World Billions," news release (Milan: 10 December 2003); deaths according to World Health Organization, cited in Christian Plumb, "WHO Says Climate Change Killing 150,000 a Year," *Reuters*, 15 December 2003.

14. José Goldemberg, "The Case for Renewable Energies," Thematic Background Paper 1, in preparation for International Conference for Renewable Energies, Bonn, Germany, 2004.

15. European Union from European Commission, "New Research Reveals the Real Costs of Electricity in Europe," press release (Brussels: 20 July 2001). Table 1 based on the following: low coal figure is for the United States, and high figure is European average, from "On Track as the Cheapest in Town," *Windpower Monthly*, January 2002, p. 30; low natural gas cost (for Europe) from David Milborrow, e-mail to author, 18 September 2002; high natural gas cost (U.S.) from U.S. Department of Energy (DOE), Office of Energy Efficiency and Renewable Energy (EREN), "Economics of BioPower," at www.eren.doe.gov/biopower/

basics/ba_econ.htm, viewed 15 July 2002; nuclear is 1993 levelized costs in California, from California Energy Commission, *1996 Energy Technology Status Report: Report Summary* (Sacramento, California: 1997), p. 73; direct-fired biomass low figure as of 1999 in the United States from Dallas Burtraw, Resources for the Future, "Testimony Before the Senate Energy and Water Development Appropriations Subcommittee," 14 September 1999; high figure for direct-fired biomass from U.S. DOE, EREN, "Biomass at a Glance," at www.eren.doe.gov/biopower/basics/index.htm, viewed 15 July 2002; hydropower low figure calculated by DOE based on 21 projects completed in 1993; high hydropower figure calculated using 30-year lifetime and real cost of capital from DOE, Energy Information Administration, *Energy Consumption and Renewable Energy Development Potential on Indian Lands* (Washington, D.C.: April 2000); photovoltaics range assumes unsubsidized, best-to-typical U.S. climates, and 6 percent interest rates, from Paul Maycock, discussion with author, 8 March 2004; wind at best sites from International Energy Agency, *Renewables for Power Generation: Status and Prospects* (Paris: IEA/Organization for Economic Development and Cooperation, 2003), p. 19; external costs from European Commission, *External Costs: Research Results on Socio-Environmental Damages Due to Electricity and Transport* (Brussels: Directorate General for Research, European Commission, 2003), p. 13.

16. Shimon Awerbuch, "Determining the Real Cost: Why Renewable Power Is More Cost-Competitive Than Previously Believed," *Renewable Energy World*, March/April 2003; Raphael Sauter and Shimon Awerbuch, "Oil Price Volatility and Economic Activity: A Survey and Literature Review," International Energy Agency Working Paper (Paris: IEA, August 2002), cited in ibid.; and Jeff Gerth, "Forecast of Rising Oil Demand Challenges Tired Saudi Fields," *New York Times*, 24 February 2004.

17. Hermann Scheer, member of German parliament, cited in Alenka Burja, "Energy Is a Driving Force for Our Civilisation: Solar Advocate," 2002, at www.foldecenter.dk/articles/Hscheer_aburja.htm, viewed 8 October 2002.

18. Brazil from José Goldemberg, Suani Teixeira Coelho, Plinio Mário Nastari, and Oswaldo Lucon, "Ethanol Learning Curve—The Brazilian Experience," *Biomass and Bioenergy* (issue number not available; accepted 14 May 2003), 2003.

19. Shares for Figure 1 calculated by Worldwatch with data from International Energy Agency (IEA), op. cit. note 8, pp. 410, 411; shares for Figure 2 calculated by Worldwatch with data from IEA, *Key World Energy Statistics 2003* (Paris: 2003).

20. Table 2 based on the following: renewables data are current use of secondary energy carriers (electricity, heat, and fuels) converted to primary energy data and adapted from José Goldemberg, ed., *World Energy Assessment: Energy and the Challenge of Sustainability* (New York: United Nations Development Programme, 2000), cited in Thomas B. Johansson, Kes McCormick, Lena Neij, and Wim Turkenburg, "The Potentials of Renewable Energy," Thematic Background Paper 10, prepared for the International Conference for Renewable Energies, Bonn, Germany, 2004, p. 3; total primary energy use for

2000 includes traditional biomass, converted from 10,089 million tons of oil equivalent, from International Energy Agency, op. cit. note 8, p. 411.

21. Transmission and distribution losses from World Bank, *World Development Report 1997* (New York: Oxford University Press, 1997), and from Indian Planning Commission, *Annual Report on the World of State Electricity Boards and Electricity Departments*, cited in M.S. Bhalla, "Transmission and Distribution Losses (Power)," in Proceedings of the National Conference on Regulation in Infrastructure Services: Progress and Way Forward (New Delhi: The Energy and Resources Institute, November 2000).

22. $100 billion invested over 10 years would be enough to make renewables competitive within 20 years, according to the Group of Eight Renewable Energy Task Force, *G8 Renewable Energy Task Force—Final Report*, cited in Organization for Economic Development and Cooperation, "Renewables: Upwardly Mobile," *OECD Observer*, 19 August 2002, at www.oecdobserver.org/news/printpage.php/aid/747/Renewables.html, viewed 10 March 2004; a single 10-percent increase in oil prices could cause economic losses in the range of hundreds of billions of U.S. dollars in International Energy Agency member countries, according to Sauter and Awerbuch, op. cit. note 16; investment modest compared to existing flows from Group of Eight, Renewable Energy Task Force, *G8 Renewable Energy Task Force—Final Report*, July 2001, pp. 44.

23. Kannappan speech at 2002 Global Windpower Conference in Paris, cited in European Wind Energy Association, "Think Paris, Act Global," *Wind Directions*, May 2002, p. 11.

24. Quote and details on Kintyre from "Wind Energy Turns Kintyre Economy Around," *Environment News Service*, 8 July 2002.

25. Jobs from Virinder Singh with BBC Research and Consulting and Jeffrey Fehrs, *The Work That Goes Into Renewable Energy*, Research Report No. 13 (Washington, D.C.: Renewable Energy Policy Project, November 2001).

26. California Public Interest Research Group, "Developing Renewable Energy Could Mean More Jobs," KTVU News, 25 June 2002, at www.bayinsider.com/partners/ktvu/news/2002/06/25_solar.html, viewed 16 July 2002; and Steve Rizer, "Davis Supports Plan To Double State's Level of Renewable-Based Electricity," *Solar & Renewable Energy Outlook*, 1 April 2002, p. 73.

27. The Energy and Resources Institute (TERI), "Survey of Energy and Environmental Situations in India," TERI report 2003IE42, 2003; and Akanksha Chaurey, TERI, e-mail to author, 16 October 2003.

28. Group of Eight Renewable Energy Task Force, op. cit. note 22, p. 9.

Technology and Market Development

1. Figure 3 based on the following: wind growth rate calculated by Worldwatch from Janet L. Sawin, "Wind Power's Rapid Growth Continues," in

Vital Signs 2003, Worldwatch Institute in Cooperation with the United Nations Environment Programme (New York: W.W. Norton & Company, 2003), pp. 38, 39; estimates for 2003 based on data from BTM Consult ApS, *World Market Update 2003* (Ringkøbing, Denmark: March 2004), and European Wind Energy Assocation and American Wind Energy Association, "Global Wind Power Growth Continues To Strengthen: Record €8 Billion Wind Power Installed in 2003," news release (Brussels/Washington, D.C.: 10 March 2004; PV data from Paul Maycock, *PV News*, various issues; fossil fuels from BP, *BP Statistical Review of World Energy* 2003 (London: 2003), and using assumed growth rates for 2003 (based on press reports) of 1 percent for coal, 1.9 percent for oil, 2 percent for natural gas; nuclear from Worldwatch Institute, *Signposts 2004*, A Worldwatch CD-ROM Resource (in preparation), with data from Worldwatch Institute Database, International Atomic Energy Agency, and press reports.

2. 100,000 MW from Eric Martinot, "The GEF Portfolio of Grid-Connected Renewable Energy: Emerging Experience and Lessons," cited in Group of Eight Renewable Energy Task Force, *G8 Renewable Energy Task Force—Final Report* (July 2001), pp. 27, 28; 300 million from Eric Martinot, Climate Change Program, Global Environment Facility, discussion with author, 4 October 2002.

3. International Energy Agency, *The Evolving Renewable Energy Market* (Paris: 1999), p. v.

4. Worldwatch estimates based on wind and PV capacity additions in 2003, assuming a 3 percent growth rate for other renewables and 2002 estimate from Eric Martinot, "Global Renewable Energy Markets and Policies," forthcoming in *New Academy Review*, Spring 2004. The calculations used Martinot's methodology, described at www.martinot.info/markets.htm (viewed 29 February 2004) and in e-mail to author, 29 February 2004.

5. €75 billion, converted using average 2003 exchange rate of €0.8854 = $1US; Renewable Energy and International Law Project, cited in "International Project To Address Barriers to Renewable Energy," *Refocus Weekly*, 14 January 2004, at www.sparksdata.co.uk/refocus/showdoc.asp?docid=1205081&accnm=1, viewed 14 February 2004.

6. Speed of progress from Intergovernmental Panel on Climate Change, Working Group 3, *Climate Change 2001: Mitigation, Summary for Policy Makers*, p. 5, at www.ipcc.ch/pub/wg3spm.pdf, viewed 10 August 2002.

7. David Milborrow, "Becoming Respectable in Serious Circles," *Windpower Monthly*, vol. 20, no. 1, January 2004, pp. 39–42.

8. Cost in early 1980s calculated by Worldwatch Institute based on Paul Gipe, "Overview of Worldwide Wind Generation," 4 May 1999, at http://rotor.fb12.tu-berlin.de/overview.html, viewed 3 March 2000; current costs from International Energy Agency, *Renewables for Power Generation: Status and Prospects* (Paris: IEA/OECD, 2003), p. 19.

9. Cheaper than gas from Milborrow, op. cit. note 7; offshore costs according to study by University of Utrecht in the Netherlands, and cited in ibid.

10. Technology trends from D.I. Page and M. Legerton, "Wind Energy Implementation During 1996," Renewable Energy Newsletter, CADDET, September 1997, at www.caddet-re.org/html/397art6.htm, viewed 22 September 1998; and from International Energy Agency, "Long-term Research and Development Needs for Wind Energy for the Time Frame 2000 to 2020," October 2001, at www.afm.dtu.dk/wind/iea, viewed 7 October 2002.

11. Average size installed worldwide in 2003 from Birger Madsen, e-mail to author, 20 February 2004.

12. Turbine sizes from Peter Fairley, "Wind Power for Pennies," *Technology Review*, July/August 2002, p. 43; and Christine Real de Azua, American Wind Energy Association, e-mail to author, 9 March 2004.

13. Small-scale turbines from "Building Integrated Wind Turbines," *RENEW: Technology for a Sustainable Future*, July/August 2002, p. 27; and William Chisholm, "Borders Homes Test New Rooftop Wind Turbines," *Scotsman*, 25 February 2004, http://business.scotsman.com/index.cfm?id=221612004, viewed 10 March 2004.

14. Sidebar 1 from the following sources: Sandia from "Sandia Strives To Help Make Energy From Less Windy Sites Cost Effective," *Solar & Renewable Energy Outlook*, vol. 29, no. 27, 15 October 2003; 20 times from "Sustainable Energy Coalition Wants Doubling of 'Clean Energy' Budget," ibid.; climatic models from Birger Madsen, BTM Consult ApS, e-mail to author, 14 September 2002; Vestas from Peter Fairley, "Wind Power for Pennies," *Technology Review*, July/August 2002, pp. 42, 43.

15. Calculated by Worldwatch with data from BTM Consult ApS, European Wind Energy Association, and American Wind Energy Association. Figure 4 from the following sources: Janet L. Sawin, "Wind Power's Rapid Growth Continues," in *Vital Signs 2003*, Worldwatch Institute in Cooperation with the United Nations Environment Programme (New York: W.W. Norton & Company, 2003), pp. 38, 39; estimates for 2003 based on data from BTM Consult ApS, *World Market Update 2003* (Ringkøbing, Denmark: March 2004), and European Wind Energy Assocation and American Wind Energy Association, "Global Wind Power Growth Continues To Strengthen: Record €8 Billion Wind Power Installed in 2003," news release (Brussels/Washington, D.C.: 10 March 2004.

16. Estimates for 2003 based on data from BTM Consult ApS, op. cit. note 15, and European Wind Energy Association and American Wind Energy Association, op. cit. note 15; number of households from EWEA and AWEA, op. cit. 15.

17. Hamburg Messe and European Wind Energy Association, "WindEnergy

Study 2004—Significant Growth for the Wind Energy Market: 110,000 Megawatts, Equivalent to 130 Billion Euros, by 2012," press release (Hamburg and Brussels: 2 March 2004).

18. "The Windicator: Operating Wind Power Capacity," *Windpower Monthly*, vol. 20, no. 1, January 2004, p. 66.

19. European Wind Energy Association, "Wind Power Expands 23% in Europe But Still Only a 3-Member States Story," news release (Brussels: 3 February 2004).

20. 2003 sales from op. cit. note 15, European Wind Energy Association and American Wind Energy Association, "Global Wind Power Growth Continues To Strengthen," and from Birger Madsen, e-mail to author, 18 February 2004; 2012 projection from Joel Makower, Ron Pernik, and Clint Wilder, "Clean Energy Trends 2003," Clean Edge Inc., February 2003, at www.cleanedge.com/reports/trends2003.pdf, viewed 3 March 2004.

21. Number employed worldwide is Worldwatch estimate based on Andreas Wagner, GE Wind Energy and European Wind Energy Association, e-mail to author, 18 September 2002; and on European Wind Energy Association, Forum for Energy and Development, and Greenpeace, *Wind Force 10* (London: 1999).

22. Birger Madsen, e-mail to author, 20 February 2004.

23. Onshore resources, estimated at 53,000 billion kilowatt-hours (kWh) (53,000 terawatthours) of electricity annually, from Michael Grubb and Niels Meyer, "Wind Energy: Resources, Systems and Regional Strategies," in Laurie Burnham, ed., *Renewable Energy Sources for Fuels and Electricity* (Washington, D.C.: Island Press, 1993), pp. 186, 187, 198; global net electricity consumption in 2000 at 15,391 billion kWh, International Energy Agency, *World Energy Outlook 2002* (Paris: 2002), p. 411.

24. Bird deaths from Paul Gipe, *Wind Power Comes of Age* (New York: John Wiley & Sons, May 1995); from National Wind Coordinating Committee, "Avian Collisions With Wind Turbines: A Summary of Existing Studies and Comparisons to Other Sources of Avian Collision Mortality in the United States," August 2001, at www.nationalwind.org/pubs/avian_collisions.pdf, viewed 3 September 2002; and from Danish Energy Agency, *Wind Power in Denmark: Technology, Policies and Results 1999* (Copenhagen: Ministry of Environment and Energy, September 1999), p. 21; mitigation from American Wind Energy Association, "Proposed Repowering May Cut Avian Deaths in Altamont," *Wind Energy Weekly*, 28 September 1998; and Real de Azua, op. cit. note 12.

25. Figure of 20 percent from R. Watson, M.C. Zinyowera, and R.H. Moss, eds., *Climate Change 1995—Impacts, Adaptations, and Mitigation of Climate Change: Scientific Technical Analyses*, Contribution of Working Group 2 to the Second Assessment Report of the IPCC (New York: Cambridge University Press, 1996); from Donald Aitken, "White Paper: Transitioning to a Renewable Energy

Future," executive summary, p. 4, prepared for the International Solar Energy Society, November 2003, at www.ises.org/shortcut.nsf/to/wp, viewed 21 March 2004; and from Strategy Unit of the Prime Minister of the United Kingdom, "Renewable Energy in the U.K.—Building for the Future of the Environment," executive summary, London, November 2001, p. 17, at www.strategy .gov.uk/files/pdf/renewanalytpap1nov.pdf, viewed 21 March 2004. Need for only minor changes from David Milborrow, *Survey of Energy Resources: Wind Energy* (London: World Energy Council, 2001), at www.worldenergy.org/ wec-geis/publications/reports/ser/wind/wind.asp, viewed 3 September 2002.

26. For information regarding wind prediction and forecasting tools and modeling, see Institut für Solare Energieversorgungstechnik (Institute for Solar Energy Technologies) website at www.iset.uni-kassel.de.

27. Makower et al., op. cit. note 20.

28. U.S. Department of Energy, National Renewable Energy Laboratory, *The Photovoltaics Promise*, NREL Report No. FS-210-24588, at www.nrel.gov/ ncpv/pdfs/24588.pdf, viewed 19 July 2002.

29. 60 percent from "Solar Power Industry Slowed by Pricey Silicon," *Reuters*, 27 January 2004; 70 percent from Paul Garvison, "PV Market Overview & Implications for Energy Storage," BP Solar, presentation dated 5 November 2003, at http://208.230.252.231/Storage%20Workshop%20Main_files%5C07Gar vison.pdf, viewed 27 February 2004.

30. Sun's energy from Richard Corkish, "A Power That's Clean and Bright," *Nature*, 18 April 2002, p. 680.

31. Calculated by Worldwatch with data from Paul Maycock, *PV News*, various issues. Figure 5 from ibid.

32. $5.2 billion conservative estimate based on global production of 742 MW from Paul Maycock, discussion with author, 23 February 2004; and estimated $7/W installation costs used by Eric Martinot, e-mail to author, 29 February 2004; job estimate based on Virinder Singh with BBC Research and Consulting and Jeffrey Fehrs, *The Work that Goes Into Renewable Energy*, Research Report No. 13 (Washington, D.C.: Renewable Energy Policy Project, November 2001), pp. 11, 12; on "Job Opportunities in Photovoltaic and Renewable Energy Engineering," at www.pv.unsw.edu.au/bepv/jobopps.htm, viewed 9 October 2002; and on 3,800 jobs for every $100 million in solar cell sales, according to the U.S. Solar Energy Industries Association.

33. 2012 projection from Makower et al., op. cit. note 20, p. 3.

34. Eric Martinot et al., "Renewable Energy Markets in Developing Countries," in *Annual Review of Energy and the Environment 2002* (Palo Alto, California: Annual Reviews), p. 3 (draft).

35. Thomas B. Johansson, Kes McCormick, Lena Neij, and Wim Turkenburg, "The Potentials of Renewable Energy," Thematic Background Paper 10, prepared for the International Conference for Renewable Energies, Bonn, Germany, 2004, p. 8.

36. Paul Maycock, discussions with author, 23 February 2004 and 8 March 2004.

37. Jesse Broehl, "Solar Energy Start-up Attacks PV Wafer Costs," SolarAccess.com, 3 December 2003. Sidebar 2 from the following sources: NREL from "High Yield Solar Cell," *RENEW: Technology for a Sustainable Future*, May/June 2002, p. 27; Spheral Solar from "ATS, Spheral Solar, Elk Enter into MOU To Design PV Roofing Products," *Solar & Renewable Energy Outlook*, vol. 29, no. 33, 1 December 2003; Evergreen Solar from "Marlboro, MA, USA: Evergreen Solar Makes Advance in String Ribbon Technology," *Solarbuzz*, 22 January 2004, at www.solarbuzz.com/News/NewsNAMA34.htm, viewed 2 February 2004, and from "Evergreen Solar Completes Research Effort for String Ribbon Technology," *Solar & Renewable Energy Outlook*, vol. 30, No. 6, 8 February 2004; Solar blinds from "Konstanz, Germany: Sunways Launches Solar Window Blinds in Europe," *Solarbuzz*, 26 January 2004, at www.solarbuzz.com/News/NewsE-UPT15.htm, viewed 2 February 2004.

38. Growth rates projected by Maycock, op. cit. note 32.

39. Drop in costs per doubling from European Photovoltaics Industry Association and Greenpeace, *Solar Generation*, October 2001, p. 14; 5 percent annual cost decline from Bernie Fischlowitz-Roberts, "Sales of Solar Cells Take Off," Eco-Economy Update (Washington, D.C.: Earth Policy Institute, 11 June 2002).

40. Joel Makower, Ron Pernick, and Andrew Friendly (Solar Catalyst Group), *Solar Opportunity Assessment Report* (Washington, D.C.: Clean Edge, Inc. and Co-op America Foundation, 2003), p.66.

41. Maycock, op. cit. note 32.

42. Generating cost range from Maycock, op. cit. note 36. Low end is for Japan, assuming 2 percent cost of money, installed costs of $5.50/W without subsidies, and 1,000 peak hours annually; high end assumes off-grid, remote, with inverters, batteries, and high installation costs.

43. Building facades from Steven Strong, "Solar Electric Buildings: PV as a Distributed Resource," *Renewable Energy World*, July-August 2002, p. 171.

44. Japan from Maycock, op. cit. note 23; California from International Energy Agency, op. cit. note 8, p. 24.

45. Manufacturers' price targets from Makower et al., op. cit. note 20.

46. Makower et al., op. cit. note 40, p. 19.

47. Maycock, op. cit. note 32.

48. Maycock, ibid.; and Makower et al., op. cit note 40, p. 19.

49. Competitive without incentives in United States by 2013 according to Navigant Consulting, "The Changing Face of Renewable Energy," Public Release Document, Burlington, MA, October 2003; and Maycock, op. cit. note 32.

50. Various estimates range from three to nine years, but most estimate about four to six; Paul Maycock, discussion with author, 8 March 2004. Also see Johansson et al., op. cit. note 35, p. 9; and International Energy Agency, op. cit. note 8, p. 62.

51. PV manufacture risks from Larry Kazmerski, "Photovoltaics—Exploding the Myths," *Renewable Energy World*, July-August 2002, p. 176, and from U.K. Department of Trade and Industry, at www.dti.gov.uk/renewable/photovoltaics.html, viewed 3 September 2002.

52. International Energy Agency, op. cit. note 8, p. 68.

53. 5 MW from "World's Largest Solar Power Station Set for Summer," *Environment News Service*, 27 January 2004; 18 MW from "World's Largest Solar Array Shapes Up," SolarAccess.com, 4 February 2004.

54. Study by International Energy Agency Task VIII, led by Kosuke Kurokawa and Kazuhiko Kato, cited in "Power from the Desert: Very Large-scale Photovoltaics," *Renewable Energy World*, May-June 2003, at www.jxj.com/magsandj/rew/2003_03/desert_power.html, viewed 27 February 2004.

55. Calculated by Worldwatch with data from BTM Consult ApS, European Wind Energy Association, and American Wind Energy Association; and nuclear data from Worldwatch Institute, *Signposts 2004*, A Worldwatch CD-ROM Resource (in preparation), with data from Worldwatch Institute Database, International Atomic Energy Agency, and press reports.

56. Wind calculated with data from BTM Consult ApS, European Wind Energy Association, and American Wind Energy Association; PV calculated with data from Paul Maycock, *PV News*, February 2004. Capacity projections assume that existing capacity remains online or is repowered. Projected total global electricity generation of 25,758 terawatthours (TWh) in 2020 from International Energy Agency, *World Energy Outlook 2002* (Paris: IEA, 2002), p. 411. Wind generation would be about 5,256 TWh annually, assuming an average capacity factor of 30 percent.

57. Wind from European Wind Energy Association and Greenpeace, *Wind Force 12: A Blueprint to Achieve 12% of the World's Electricity From Wind Power by 2020* (Brussels and Amsterdam, May 2003), p. 6; PV projection from European Photovoltaics Industry Association (EPVA) and Greenpeace, op. cit. note 39, p. 5.

Two Success Stories: Germany and Japan

1. Andreas Wagner, GE Wind and European Wind Energy Association, discussion with author, 10 September 2002.

2. Utilities from Gerhard Gerdes, Deutches Windenergie Institut, discussion with author, 7 December 2000.

3. Grid from Johannes Lackmann, "The German Market Promotion Initiatives for Renewable Energies," Bundesverband Erneuerbare Energie e.V. (German Wind Energy Association), presentation at conference of the World Council for Renewable Energy, 2002, at www.world-council-for-renew able-energy.org/downloads/WCRE_Lackmann.pdf, viewed 15 November 2003.

4. Jochen Twele, Bundesverband WindEnergie e.V. (German Wind Energy Association), discussions with author, April 1999, 5 December 2000.

5. Investment amount from Andreas Wagner, GE Wind and European Wind Energy Association, e-mail to author, 18 September 2002.

6. Roland Mayer, Bundesministerium für Wirtschaft (German Federal Ministry of Economics), e-mail to author, 30 March 2001; Deutches Windenergie Institut, *Wind Energy Information Brochure* (Wilhelmshaven, Germany: 1998); and German Ministry for the Environment, Nature Conservation, and Nuclear Safety (Deutsche Bundesministerium für Umwelt, Naturschutz, und Reaktorsicherheit), *Environmental Policy: The Federal Government's Decision of 29 September 1994 on Reducing Emissions of CO_2, and Emissions of Other Greenhouse Gases, in the Federal Republic of Germany* (Bonn: November 1994).

7. Wind from Andreas Wagner, GE Wind and European Wind Energy Association, e-mails to author, 10 September and 18 September 2002, and discussion with author 10 September 2002; PV from Ingrid Weiss and Peter Sprau, "100,000 Roofs and 99 Pfennig—Germany's PV Financing Schemes and the Market," *Renewable Energy World*, January-February 2002.

8. 1990 from Bundesverband WindEnergie e.V. (German Wind Energy Association), "Installationszahlen in Deutschland, 1988–Ende 2000," at www.wind-energie.de/statistic/Deutschland.html, viewed 14 March 2001; 2003 from Deutsches Windenergie-Institut, "Windenergie—Mehr Leistung Neu Installiert als Erwartet," press release (Berlin: 27 January 2004). Figure 6 from the following sources: Germany from Bundesverband WindEnergie, op. cit. this note, and European Wind Energy Association (EWEA), various press releases; Spain from Instituto para la Diversificación y Ahorro Energético (Institute for Energy Diversification and Savings), Energía Hidroeléctrica de Navarra (Hydroelectric Energy of Navarra), and Asociación de Productores de Energías Renovables (Association of Renewable Energy Producers) data supplied by José Santamarta, e-mail to author, 19 October 2002; and EWEA, op. cit. this note; U.S. from Paul Gipe, discussions with author, and faxes, 1 October 1998 and 23 March 2001; and American Wind Energy Association, various press releases; and some data for all countries from BTM Consult ApS, *World Market Update*, various years.

9. 2003 share based on average wind year and from Ralf Bischof, German Renewable Energy Association, e-mail dated 26 February 2004, with data from Deutsches Windenergie-Institut; 2001 share from "German 2002 Wind Power Market Up 22 Pct," *Reuters*, 24 January 2003.

10. Wind's share in Schleswig-Holstein from C. Ender, Deutsches WindEnergie-Institut, "Wind Energy Use in Germany—Status 30.06.03," *DEWI Magazin* no. 23, August 2003.

11. Growth rate and capacity calculated with data from International Energy Agency, Photovoltaic Power Systems Program, "Statistics by Country," at www.oja-services.nl/iea-pvps/statistics/countries.htm, viewed 16 March 2004; and Paul Maycock, discussion with author, 8 March 2004.

12. Expected PV manufacturing expansion from Reiner Gärtner, "Fatherland and Sun," *Red Herring*, 22 July 2002; and from International Energy Agency, Photovoltaic Power Systems Programme, "Germany—National Status Report 2002: Production of Photovoltaic Cells and Modules," 27 August 2003, at www.oja-services.nl/iea-pvps/countries/germany/index.htm, viewed 27 March 2004.

13. €9.6 billion, converted using average 2003 exchange rate of €0.8854 = $1U.S.; German Federal Ministry for the Environment, Nature Conservation, and Nuclear Safety, "Renewable Energies: The Way Forward," Berlin, November 2003, p. 5.

14. Bundesverband Windenergie, "German Wind Power Still Flying High," press release (Osnabrück: 22 February 2003).

15. €24,000, converted using average 2003 exchange rate of €0.8854 = $1U.S. "Germany Reaches 100,000 Roof Milestone—But Faces Meltdown," *Platts Renewable Energy Report*, Issue 53, July 2003.

16. Biogas share was 35.5 percent in 2001, from Ian French, "Biogas in Europe: Huge Potential in a Growing Market," *Renewable Energy World*, vol. 6, no. 4, July-August 2003, pp. 120–131; solar thermal from "Trends in European Solar Thermal Market," SolarAccess.com, 11 July 2003.

17. CO_2 reductions from "German Wind Generation to Rise 25 pct in 2002—Firms," *Reuters*, 5 September 2002.

18. Dieter Uh, Secretariat, International Conference for Renewable Energies 2004, e-mails to author 2 December 2003, and 8 January 2004.

19. Wind share target announced by German Environment Minister Jürgen Trittin and cited in European Wind Energy Association, "Another Record

Year for European Wind Power," news release (Brussels: 20 February 2002); off-shore target from interview with Jürgen Trittin, in *Wind Directions*, January-February 2004, pp. 29–30.

20. German Federal Ministry for the Environment, Nature Conservation and Nuclear Safety, "Renewable Energies: The Way Forward," Berlin, November 2003, p. 7.

21. New technologies include wind, PV, and fuel cells; costs estimated from data presented by Winfried Hoffmann, RWE SCHOTT Solar GmbH, "PV Solar Electricity: One Among the New Millennium Industries—Industry Political Actions for a Sustainable European PV Industry," presentation given for European Photovoltaic Association, Brussels, 26 August 2003.

22. Half the insolation from Paul Maycock, discussion with author, 23 February 2004; capacity relative to United States calculated by Worldwatch with data from International Energy Agency, Photovoltaic Power Systems Programme, op. cit. note 11; Paul Maycock, *PV News*, various issues; and Maycock, op. cit. note 11.

23. International Energy Agency, Photovoltaic Power Systems Programme, "Japan—National Status Report 2002: Framework for Deployment—Nontechnical Factors," August 2003, at www.oja-services.nl/iea.pvps/nsr02/jpn.htm, viewed 25 November 2003.

24. Capacity from International Energy Agency, Photovoltaic Power Systems Programme, "Statistics by Country," at www.oja-services.nl/iea-pvps/statistics/countries.htm, viewed 16 March 2004; share of manufacture calculated by Worldwatch with data from Paul Maycock, *PV News*, various issues.

25. Seven-fold increase and 1994 budget from Curtis Moore and Jack Ihle, "Renewable Energy Policy Outside the United States," Issue Brief No. 14, Renewable Energy Policy Project, Washington, D.C., 1999; 2001 and 2004 budgets from Paul Maycock, *PV News*, vol. 23, no. 2, February 2004, p. 2.

26. International Energy Agency, op. cit. note 23.

27. Ken-Ichiro Ogawa, New Energy and Industrial Technology Development Organization, "Japan: PV technology status and prospects," 24 November 2002, at www.oja-services.nl/iea-pvps/ar00/jpn.htm, viewed 21 November 2003.

28. International Energy Agency, op. cit. note 23.

29. Figure 7 from the following sources: Paul Maycock, *PV News*, various issues; and Maycock, op. cit. note 11.

30. Paul Maycock, *PV News*, July 2003, p. 2.

31. Maycock, op. cit. note 22, discussion with author.

32. Paul Maycock, "PV market update," *Renewable Energy World*, vol. 6, no. 4, July-August 2003, pp. 84–101, and Maycock, op. cit. note 30.

33. Number of households from International Energy Agency, op. cit. note 23; amount of subsidy from Maycock, op. cit. note 30.

34. Growth rate and capacity calculated with data from International Energy Agency, Photovoltaic Power Systems Programme, op. cit. note 11; and Maycock, op. cit. note 11.

35. Tim Sharp, "'New Energy' for Japan: Government Sets Targets for 2010," *Cogeneration and On-Site Power Production*, vol. 4, issue 5, September-October 2003, http://jxj.com/magsandj/cospp/2003_05/new_energy.html.

36. Maycock, op. cit. note 22, discussion with author.

37. 80 percent drop in costs for 3-kW systems estimated from graphs in Doug Allday, "KYOCERA Solar, Inc.," presentation for UPEx Conference, 9 October 2003, www.solarelectricpower.ewebeditpro/items/O63F3392.pdf, viewed 27 February 2004; global module costs declined an estimated 45 percent between 1993 and 2002, based on graph with data from Strategies Unlimited and BP Solar internal estimates, in Paul Garvison, "PV Market Overview & Implications for Energy Storage," BP Solar, presentation dated 5 November 2003, at http://208.230.252.231/Storage%20Workshop%20Main_files%5C07Garvison.pdf, viewed 27 February 2004.

38. Maycock, op. cit. note 22,m discussion with author.

39. Cost depends on system costs; low end assumes 2 percent cost of money. Maycock, op. cit. note 11.

40. 2002 increase from Paul Maycock, "PV market update," *Renewable Energy World*, vol. 6, no. 4, July/August 2003, pp. 84–101; 2003 increase calculated with data from Paul Maycock, *PV News*, March 2004.

41. Share of production and Sharp calculated with data from Paul Maycock, *PV News*, March 2004.

Policy Lessons From Around the World

1. Year-end 1993 capacity (52 MW) from Instituto para la Diversificación y Ahorro Energético, Spain; 2000 from BTM Consult ApS, *World Market Update 2002* (Ringkøbing, Denmark: March 2003); and European Wind Energy Association, news releases.

2. Wind's share of 2003 power generation is calculated by Worldwatch using official data, National Commission on Energy, Government of Spain, January 2004; manufacturers from BTM Consult ApS, op. cit. note 1, p. 15.

3. American Wind Energy Association, "Global Wind Energy Market Report:

Wind Energy Industry Grows at Steady Pace, Adds Over 8,000 MW in 2003," p. 3, at www.awea.org/pubs/documents/globalmarket2004.pdf, viewed 19 March 2004.

4. American Wind Energy Association, "Texas Wind Energy Development," 19 June 2002, at www.awea.org/projects/texas.html, viewed 24 July 2002.

5. Number of states with Renewables Portfolio Standard laws from U.S. Department of Energy, Office of Energy Efficiency and Renewable Energy, "California Mandates 20 Percent Renewable Power by 2017," at www.eren.doe.gov/news/news_detail.cfm?news_id=325, viewed 25 September 2002.

6. Problems with U.K. law from British Wind Energy Association, "Promoting Wind Energy in and Around the U.K.—The Government's Policy for Renewables, NFFO, and the Fossil Fuel Levy," at www.bwea.com/ref/nffo.html, viewed 3 September 2002; problems for small firms and cooperatives from Wilson Rickerson, "Germany and the European Wind Energy Market," (Berlin: Bundesverband Erneuerbare Energie e.V., 2002); p. 4.

7. 1999 statistics from World Energy Council, *Survey of Energy Resources: Wind Energy* (London: 2001).

8. U.K. capacity from European Wind Energy Association, "Wind Power Expands 23% in Europe But Still Only a 3-Member State Story," news release (Brussels: 3 February 2004).

9. Database of State Incentives for Renewable Energy, University of North Carolina, www.dsireusa.org/summarytables/reg1.cfm?&CurrentPageID=7, viewed 25 January 2004.

10. Share of EU total from European Wind Energy Association, cited in "Wind Power Grows by 23% in Europe in 2003," *Wind Energy Weekly*, vol. 23, no. 1079, 13 February 2004; share of global total calculated by Worldwatch with ibid., and BTM Consult ApS, *World Market Update 2003* (Ringkøbing, Denmark, March 2004). Denmark's pricing law has been terminated, but some wind farms still operate under its terms.

11. One study estimates that the price increase for electricity consumers caused by the pricing law was only 0.11 eurocent/kWh in 2000, and will be 0.19 eurocent/kWh in 10 years assuming a doubling of renewables' share of total generation; see Johannes Lackmann, "The German Market Promotion Initiatives for Renewable Energies," Bundesverband Erneuerbare Energie e.V., presentation at conference of the World Council for Renewable Energy, 2002, at www.world-council-for-renewable-energy.org/downloads/WCRE_Lack mann.pdf, viewed 15 November 2003. Another estimate puts additional costs at 0.26 cents/kWh in 2001; Dieter Uh, thematic advisor, Secretariat, International Conference for Renewable Energies, e-mails to author, 2 December 2003 and 8 January 2004.

12. Janet L. Sawin, "The Role of Government in the Development and Diffusion of Renewable Energy Technologies: Wind Power in the United States, California, Denmark, and Germany, 1970–2000" (dissertation, The Fletcher School, Tufts University), September 2001 (Ann Arbor, MI: UMI, 2001), pp. 204, 205.

13. Untested designs from Alfred J. Cavallo, Susan M. Hock, and Don. R. Smith, "Wind Energy: Technology and Economics," in Laurie Burnham, ed., *Renewable Energy Sources for Fuels and Electricity* (Washington, D.C.: Island Press, 1993), p. 150; lack of generation from Alan J. Cox et al., "Wind Power in California: A Case Study of Targeted Tax Subsidies," in Richard J. Gilbert, ed., *Regulatory Choices: A Perspective on Developments in Energy Policy* (Berkeley: University of California Press, 1991), p. 349; and from Vincent Schwent, California Energy Commission, discussion with author, 6 May 1999.

14. Indian Ministry of Non-Conventional Energy Sources, *Annual Report 1999–2000*, at http://mnes.nic.in/frame.htm?publications.htm, viewed 29 July 2002.

15. Lower capacity factors and some nonfunctioning turbines from Eric Martinot et al, "Renewable Energy Markets in Developing Countries," in *Annual Review of Energy and the Environment 2002* (Palo Alto, California: Annual Reviews), pp. 11, 20 (draft).

16. Mark Clayton, "Solar Power Hits Suburbia," *Christian Science Monitor*, 12 February 2004.

17. California from ibid. Sidebar 3 from the following sources: CESA from "Public Funds in 12 States Create Joint Effort to Promote Renewables," *Wind Energy Weekly*, vol. 23, no. 1077, 30 January 2004; California payments and rebates and third-largest market from California Energy Commission, Sacramento, California, 2003, www.energy.ca.gov/renewables, viewed 13 December 2003; 50-percent cost reduction for PV from "Solar Electric Systems Boost California Power Grid," *SolarAccess.com*, 5 November 2003; number of companies from Celia Lamb, "New Laws Reshape, Fuel Solar Power Growth," *Sacramento Business Journal*, 20 October 2003; solar bond initiative from Vote Solar Initiative, San Francisco, California, www.votesolar.org, 2 December 2003; and Kendra Mayfield, "'Fog City' Catches a Few Rays," PowerLight Corporation, news release, San Francisco, 7 January 2003).

18. Benefits to those with higher income from Sawin, op. cit. note 12, p. 151.

19. Janet L. Sawin, "Charting a New Energy Future," in Linda B. Starke, ed., *State of the World 2003* (New York: W.W. Norton & Company, 2003), p. 102.

20. Sawin, op. cit. note 12, pp. 151, 340, 341.

21. Michael Eckhart, Jack Stone, and Keith Rutledge, "Financing PV Growth: Why Finance May Be the Key to Real Expansion," *Renewable Energy World*, vol.

6, no. 3, May-June 2003, pp. 112–127.

22. Eric Martinot et al., op. cit. note 15, pp. 8, 22.

23. Indian loans from Indian Ministry of Non-Conventional Energy Sources, op. cit. note 14, p. 53; funding in India from "Why Renewables Cannot Penetrate the Market," *Down to Earth*, 30 April 2002, p. 35.

24. Tax credit in Arkansas from Robert Righter, *Wind Energy in America: A History* (Norman: University of Oklahoma Press, 1996), p. 205; wind's potential share of Arkansas' electricity calculated by Worldwatch with consumption data from U.S. Department of Energy, Energy Information Administration, at www.eia.doe.gov/cneaf/electricity/st_profiles/arkansas/ar.html, viewed 7 September 2002; and with wind potential from Battelle/Pacific Northwest Laboratory (PNL), *Assessment of Available Windy Land Area and Wind Energy Potential in the Contiguous United States* (Battelle/PNL, August 1991), cited in Jan Hamrin and Nancy Rader, *Investing in the Future: A Regulator's Guide to Renewables* (Washington, D.C.: National Association of Regulatory Utility Commissioners, February 1993), p. A-11.

25. Sawin, op. cit. note 12, chapter 4.

26. Li Hua, "China's Solar Thermal Industry: Threat or Opportunity for European Companies?" *Renewable Energy World*, vol. 5, no. 4, July-August 2002, p. 107.

27. Austria from Larry Goldstein, John Mortensen, and David Trickett, "Grid-Connected Renewable-Electric Policies in the European Union," National Renewable Energy Laboratory, NREL/TP.620.26247, May 1999; Germany from Bundesministerium für Umwelt, Naturschutz, und Reaktorsicherheit, *Environmental Policy: The Federal Government's Decision of 29 September 1994 on Reducing Emissions of CO2, and Emissions of Other Greenhouse Gases, in the Federal Republic of Germany* (Bonn: November 1994).

28. Indian programs from Indian Ministry of Non-Conventional Energy Sources, op. cit. note 14.

29. Daniel M. Kammen, "Bringing Power to the People: Promoting Appropriate Energy Technologies in the Developing World," *Environment*, June 1999.

30. European Actions for Renewable Energy (PREDAC), "4 Good Reasons to Favour Local Investment," (undated—2002 or 2003), at www.cler.org/predac/wp1, viewed 18 December 2003.

31. Benefits of cooperatives from Sawin, op. cit. note 12, p. 377.

32. "'Minigrids' Solve South Asia Power Crisis," *BBC News*, 27 October 2003.

33. Dominance of Danish turbine manufacturers from Søren Krohn, "Danish Wind Turbines: An Industrial Success Story," 21 January 2000, at www.wind power.dk/articles/success.htm, viewed 28 January 2000; and from Birger Madsen, BTM Consult, discussion with author, 8 December 2000, and e-mail to author, 14 September 2002.

34. Sawin, op. cit. note 12, p. 375.

35. United Kingdom from Madsen, op. cit. note 33.

36. "Ken Gives Green Light to Solar," Solar Century, news release (London: 26 January 2004); and "London Mayor Launches Renewables Program," SolarAccess.com, 4 March 2004.

Changing Government Approaches to Energy Policy

1.   José Goldemberg, ed., *World Energy Assessment: Energy and the Challenge of Sustainability* (New York: UN Development Programme, 2000).

2.   Howard Geller, *Energy Revolution: Policies for a Sustainable Future* (Washington, D.C.: Island Press, 2003).

3.   Eric Martinot, Climate Change Program, Global Environment Facility, e-mail to author, 9 October 2002.

4.   "The World Bank and Fossil Fuels: At the Crossroads," Sustainable Energy and Economy Network, Institute for Policy Studies Brief (Washington, D.C.: September 2003), at www.seen.org/pages/reports/WB_brief_0903.shtml, viewed 22 February 2004.

5.   Annual investments in energy infrastructure from José Goldemberg, "Rural Energy in Developing Countries," in Goldemberg, op. cit. note 1, p. 348; and from Eric Martinot, Climate Change Program, Global Environment Facility, e-mail to author, 9 October 2002.

6.   John Perlin, "Electrifying the Unelectrified," *Solar Today*, November/December 1999.

7.   International Energy Agency, *World Energy Investment Outlook* (Paris: 2003).

8.   Manufacturers' fears from California Energy Commission, *Wind Energy Program Progress Report* (Sacramento: 1982), p. 23.

9.   India from "Renewables Deserted?" *Down to Earth*, 30 April 2002, from Indian Ministry of Non-Conventional Energy Sources, *Annual Report 1999-2000*, at http://mnes.nic.in/frame.htm?publications.htm, viewed 29 July 2002, p. 69; and from "Why Renewables Cannot Penetrate the Market," *Down to Earth*, 30 April 2002, p. 33.

10. Torgny Møller, "Government closes door in Denmark," *Windpower Monthly*, July 2002, and Birger Madsen, e-mail to author, 11 March 2004; and jobs from ibid.

11. Risk from Janet L. Sawin, "The Role of Government in the Development and Diffusion of Renewable Energy Technologies: Wind Power in the United States, California, Denmark, and Germany, 1970–2000," (dissertation, The Fletcher School, Tufts University), September 2001 (Ann Arbor, MI: UMI, 2001), pp. 360–63, 379; and reappropriation, etc. from Dieter Uh, Secretariat, International Conference for Renewable Energies 2004, e-mails to author, 2 December 2003 and 8 January 2004.

Unlocking Our Energy Future

1. Distance from Mars on date of *Spirit's* landing from Mars Institute, "Mars Exploration Rovers—Quick Facts," February 2004, www.marsinstitute.info/epo/merfacts.html, viewed 12 March 2004.

2. GE share of wind market from "Wind Power Study: Slow Growth, Consolidation," SolarAccess.com, 3 February 2004; GE in PV market from "AstroPower Inc. Strikes Deal to Sell Certain Business Assets to GE Energy," *Solar & Renewable Energy Outlook*, vol. 30, no. 7, 15 February 2004, p. 37; and Barnaby J. Feder, "GE Signals a Growing Interest in Solar, *New York Times*, 13 March 2004.

3. Citigroup, "Corporate Citizenship: The Environment—Citigroup's New Environmental Initiatives," January 2004, at www.citigroup.com/citigroup/environment/initiatives.htm, viewed 2 March 2004.

4. Fourfold increase based on Worldwatch estimate that global renewable investments totaled about $20.3 billion in 2003 (see note 1, Introduction) and projected market value of €75 billion (converted using average 2003 exchange rate of €0.8854=$1U.S.) from Renewable Energy and International Law Project, cited in "International Project To Address Barriers to Renewable Energy," *Refocus Weekly*, 14 January 2004, at www.sparksdata.co.uk/refocus/showdoc.asp?docid=1205081&accnm=1, viewed 1 March 2004.

5. Prime Minister Tony Blair, speech entitled "Environment: The Next Steps," *Reuters*, 6 March 2001, cited in Group of Eight, Renewable Energy Task Force, *G8 Renewable Energy Task Force—Final Report*, July 2001, p. 16.

6. European Union goal under Renewables Directive from European Wind Energy Association, "European Renewable Electricity Directive: The Final Version," *Wind Directions*, January 2002, pp. 10, 11.

7. China wind from Yang Jianxiang, "Large Scale Market Closer in China," *Windpower Monthly*, vol. 20, no. 1, January 2004, pp. 56–58; China PV from "Chinese Government to Boost Solar Investments," SolarAccess.com, 12 January 2004.

8. Martinot, op. cit. note 4.

9. Brazil from Suani T. Coelho, executive assistant for the Secretary of State for the Environment, São Paulo, Brazil, discussion with author, 25 July 2002.

Table 3 from the following sources: California from Mark Glyde, "California Passes RPS," *NW Energy Coalition Report*, vol. 21, no. 8, 2002 September, p. 2; China from Yang Jianxiang, op. cit. note 7, and European Wind Energy Association and American Wind Energy Association, "Global Wind Power Growth Continues To Strengthen: Record € 8 Billion Wind Power Installed in 2003," news release (Brussels/Washington, D.C.: 10 March 2004; EU targets from European Union goal from European Wind Energy Association, op. cit. note 6; EU recent totals from "Progress Report on the Implementation of the European Renewables Directive," World Wildlife Fund (World Wide Fund for Nature), October 2003, p. 9, at www.panda.org/downloads/europe/renew ablesdirectiveoctober2003.pdf, viewed 18 March 2004; and European Commission, "Energy for the Future: Renewable Sources of Energy. White Paper for a Community Strategy and Action Plan," Brussels, 26 November 2003; Germany target from "Berlin Provides Focus for European Vanguard," *Wind Directions*, January-February 2004, p. 25; Germany recent total from "Renewable Energy Generation in Germany," Volker-Quaschning, www .volker-quaschning.de/datserv/ren-Strom-D/index_e.html, viewed 13 March 2004; Japan from Tim Sharp, "'New Energy' for Japan: Government Sets Targets for 2010," *Cogeneration and On-Site Power Production*, vol. 4, issue 5, September-October 2003, http://jxj.com/magsandj/cospp/2003_05/new _energy.html, viewed 13 December 2003; International Energy Agency, Photovoltaic Power Systems Programme, 2003, www.iea-pvps.org, viewed 13 December 2003; and Paul Maycock, discussion with author, 8 March 2004; Latin America and Caribbean from "Renewables Move Up the International Agenda," *Wind Directions*, January-February 2004, pp. 18, 19; Spain from "Navarra: Moving Towards 100% Renewables," *Wind Directions*, January-February 2004, p. 26; Thailand from International Institute for Energy Conservation, "Transitioning to Renewable Energy: Policies, Programs and Institutions That Support Long-term Transformation of Emerging Renewable Energy Markets," white paper written for the Böll Foundation (Draft), Washington, D.C., December 2003, p. 34.

10. There were 255 House and Senate members as of 12 February 2004, according to Common Dreams, "78 Groups Urge Congressmembers To Join Sustainable Energy Caucus," press release (Washington, D.C.: 12 February 2004).

11. U.S. states from Interstate Renewable Energy Council, Database of State Incentives for Renewable Energy, at www.dsireusa.org, viewed 14 October 2002; governors from "Oregon Gov. Plans for Renewable Energy," SolarAccess.com, 23 January 2004.

12. Paul Appleby, BP Solarex, United Kingdom, cited in Greenpeace, *Breaking the Solar Impasse* (Amsterdam: September 1999), p. 2.

13. G8 Renewable Energy Task Force, op. cit. note 5, p. 9.

14. BP from Amanda Griscom, "Got Sun? Marketing the Revolution in Clean Energy," *Grist Magazine*, 29 August 2002; Shell from Simon Tuck, "Royal Dutch/Shell Taking Minority Stake in Iogen," *Globe & Mail*, at www.gogreen

industries.com/Clippings/RoyalDutchShell8May02.pdf, viewed 10 October 2002; and from Karen De Segundo, Shell Renewables, cited in "Renewables Could Supply Up to One-Third of World Energy by 2050," *Refocus Weekly*, 12 November 2003.

15. David Jones, Shell WindEnergy, in *Platts Global Energy*, 2001, at www.platts.com/renewables/investment.shtml, viewed 10 October 2002.

16. Meeting U.S. needs with wind calculated with data from Batelle/Pacific Northwest Laboratory, *Assessment of Available Windy Land Area and Wind Energy Potential in the Contiguous United States* (Batelle/PNL, August 1991), cited in Jan Hamrin and Nancy Rader, *Investing in the Future: A Regulator's Guide to Renewables* (Washington, D.C.: National Association of Regulatory Utility Commissioners, February 1993), p. A-11; solar in Nevada is assuming parabolic trough systems, from U.S. Department of Energy, "Concentrating Solar Power Technologies Overview," at www.energylan.sandia.gov/sunlab/overview.htm, viewed 25 January 2002.

17. Hydrogen cars from Vijay V. Vaitheeswaran, *Power to the People* (New York: Farrar, Straus and Giroux, 2003), p. 15; and Matt Daily, "Dow, GM Launch Largest Commercial Fuel Cell," *Reuters*, 12 February 2004.

18. "New Analysis Projects 20% Renewables by 2020," *Wind Directions*, January-February 2004, pp. 33–35.

19. Based on results of LTI-Research Group study begun in 1998, from "Olav Hohmeyer, "Switching the European Economy to Renewables Over the Next 50 Years," *RENEW*, Natta Newsletter 147, January-February 2004.

20. Intergovernmental Panel on Climate Change, *Climate Change 2001: The Scientific Basis* (Cambridge, U.K.: Cambridge University Press, 2001), p. 12.

Index

Other Worldwatch Papers

On Climate Change, Energy, and Materials

On Ecological and Human Health

On Economics, Institutions, and Security

On Food, Water, Population, and Urbanization

Other Publications from the Worldwatch Institute

State of the World 2004   NOW AVAILABLE!
Worldwatch's flagship annual is used by government officials, corporate planners, journalists, development specialists, professors, students, and concerned citizens in over 120 countries. Published in more than 20 different languages, it is one of the most widely used resources for analysis.

Signposts 2003
This CD-ROM provides instant, searchable access to over 1365 pages of full text from the last three editions of *State of the World* and *Vital Signs*, comprehensive datasets going back as far as 50 years, various historical timelines, and easy-to-understand graphs and tables. Fully indexed, *Signposts 2003* contains a powerful search engine for effortless search and retrieval. Plus, it is platform independent and fully compatible with all Windows (3.1 and up), Macintosh, and Unix/Linux operating systems.

Vital Signs 2003
Written by Worldwatch's team of researchers, this resource provides comprehensive, user-friendly information on key trends and includes tables and graphs that help readers assess the developments that are changing their lives for better or for worse.

State of the World Library 2004
Subscribe to the State of the World Library and join thousands of decisionmakers and concerned citizens who use these publications to stay current on emerging environmental issues. Sate of the World Library 2004 includes Worldwatch's flagship annual, *State of the World 2004,* plus three other timely publications: *Mainstreaming Renewable Energy in the 21st Century; Eat Here: Homegrown Pleasures in a Global Supermarket;* and *The Security Demographic.*

World Watch
This award-winning bimonthly magazine is internationally recognized for the clarity and comprehensiveness of its articles on global trends. Keep up-to-speed on the latest developments in population growth, climate change, species extinction, and the rise of new forms of human behavior and governance.

To make a tax-deductible contribution or to order any of Worldwatch's publications, call us toll-free at 888-544-2303 (or 570-320-2076 outside the U.S.), fax us at 570-320-2079, e-mail us at wwpub@worldwatch.org, or visit our website at www.worldwatch.org.